浙江师范大学"化学"省级实验教学示范中心重点建设项目资助

化工原理
实验及仿真（汉英对照）

Experiment and Simulation of Chemical Engineering Principle

■ 主编 代伟　■ 参编 滕波涛　汤岑　刘亚　胡鑫　蒋永福

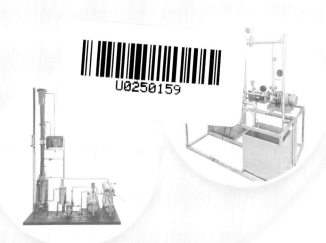

武汉大学出版社

图书在版编目(CIP)数据

化工原理实验及仿真:汉英对照/代伟主编. —武汉:武汉大学出版社,2018.4
ISBN 978-7-307-19973-6

Ⅰ.化… Ⅱ.代… Ⅲ.化工原理—实验—汉、英 Ⅳ.TQ02-33

中国版本图书馆 CIP 数据核字(2017)第 329058 号

责任编辑:鲍 玲　　责任校对:汪欣怡　　版式设计:马 佳

出版发行:**武汉大学出版社**　(430072　武昌　珞珈山)
　　　　　(电子邮件:cbs22@whu.edu.cn　网址:www.wdp.com.cn)
印刷:湖北睿智印务有限公司
开本:787×1092　1/16　印张:10　字数:250 千字　插页:1
版次:2018 年 4 月第 1 版　2018 年 4 月第 1 次印刷
ISBN 978-7-307-19973-6　　定价:29.00 元

版权所有,不得翻印;凡购我社的图书,如有质量问题,请与当地图书销售部门联系调换。

前　言

本书是根据浙江师范大学应用化学学科教师多年的化工原理实验教学实践经验编写而成的。本书在《化工原理实验》讲义的基础上，参考了国内外相关教材及文献，增加了化工原理仿真实验和英文对照两部分内容。

全书根据化工过程中经典的"三传"，分为动量传递实验、热量传递实验和质量传递实验三个章节。具体包括离心泵特性曲线的测定、管路流体阻力的测定、流量计校核实验、"空气-水"气液传热实验、"水-蒸汽"给热系数实验、"乙醇-水"筛板精馏实验、填料吸收塔、流化床干燥实验八个典型化工原理实验。本书注重理论与实践相结合，注重学生工程观点的培养和实践能力的提高，可供高等院校尤其是开设双语及仿真教学的应用化学及相近专业师生参考，亦可供从事化工实验研究的研究生和实验人员参考。

本书由浙江师范大学应用化学学科代伟老师主编，滕波涛、汤岑、刘亚、胡鑫、蒋永福等老师参编，此外感谢北京东方仿真软件技术有限公司和浙江中控技术股份有限公司在设备和软件方面提供的技术支撑和帮助。

由于编者水平和能力有限，本书不妥之处在所难免，恳请读者提出宝贵意见。

编　者
浙江师范大学
2017 年 12 月

Preface

This book was compiled on the basis of years of teaching practices of the chemical engineering principle experiments by the teachers in Department of Applied Chemistry, Zhejiang Normal University. This book which is based on the precedent handouts *Experiments of Chemical Engineering* has added the contents of simulation experiments of chemical engineering as well as the English version.

The whole book, according to the classical "Three transfer" theory in chemical engineering, is mainly divided into three sections: the momentum transfer, mass transfer and heat transfer. In detail, it contains eight typical experiments of chemical engineering as following: *the measurement of the characteristic curves of centrifugal pump, the measurement of pipeline flow resistance, the graduation of flowmeter, the gas-liquid heat transfer experiment in the system of air and water, the measurement of the thermal coefficient of heat steam, the distillation experiment of ethanol-water system in sieve distillation column, the absorption experiment in a packing tower, and the desiccation experiment in a fluidized bed*. The book aims at the combination of doctrine and practice, emphasizing the cultivation of engineering perspectives along with practical abilities of students; It is supportive for the use as a reference by teachers and students in other universities especially those who have similar majors as applied chemistry or employ bilingual and simulation teaching.

This book is compiled by the chief editor Dai Wei, and edited by Teng Botao, Tang Cen, Liu Ya, Hu Xin, Jiang Yongfu and other teachers. Furthermore, thanks for the technical supporting and assistance in devices and software by Beijing Orient Simulation Software Technology Co., Ltd. and Zhejiang Central Control Technology Limited by Share Ltd.

Limited by the abilities of the editors, the book inevitably contains inadequacies. We earnestly welcome valuable advices from readers!

Editor
December, 2017.

目　录

第一章　典型动量传递实验 ·· 1
　1.1　离心泵特性曲线的测定 ·· 1
　　1.1.1　实验目的 ·· 1
　　1.1.2　实验原理 ·· 1
　　1.1.3　实验装置 ·· 2
　　1.1.4　实验仿真装置 ·· 2
　　1.1.5　实验步骤 ·· 2
　　1.1.6　仿真实验步骤 ·· 4
　　1.1.7　数据记录及处理 ··· 6
　　1.1.8　仿真实验记录及处理 ··· 7
　　1.1.9　问题讨论 ·· 10
　1.2　管路流体阻力的测定 ·· 10
　　1.2.1　实验目的 ·· 10
　　1.2.2　实验原理 ·· 10
　　1.2.3　实验装置 ·· 12
　　1.2.4　实验仿真装置 ·· 13
　　1.2.5　实验步骤 ·· 13
　　1.2.6　仿真实验步骤 ·· 15
　　1.2.7　数据记录及处理 ··· 18
　　1.2.8　仿真实验记录及处理 ··· 19
　　1.2.9　问题讨论 ·· 20
　1.3　流量计校核实验 ··· 20
　　1.3.1　实验目的 ·· 20
　　1.3.2　实验原理 ·· 21
　　1.3.3　实验装置 ·· 23
　　1.3.4　实验仿真装置 ·· 24
　　1.3.5　实验步骤 ·· 24
　　1.3.6　仿真实验步骤 ·· 25
　　1.3.7　数据记录及处理 ··· 28

1.3.8　仿真实验记录及处理 …………………………………………… 28
　　1.3.9　问题讨论 ……………………………………………………… 30

第二章　典型热量传递实验 …………………………………………………… 31
2.1　"空气-水"气液传热实验 ……………………………………………… 31
　　2.1.1　实验目的 ……………………………………………………… 31
　　2.1.2　实验原理 ……………………………………………………… 31
　　2.1.3　实验装置 ……………………………………………………… 31
　　2.1.4　实验步骤 ……………………………………………………… 33
　　2.1.5　数据记录与结果处理 …………………………………………… 33
　　2.1.6　虚拟仿真实验设备流程 ………………………………………… 33
　　2.1.7　换热虚拟仿真实验操作 ………………………………………… 34
　　2.1.8　虚拟仿真数据处理 ……………………………………………… 37
　　2.1.9　思考题 ………………………………………………………… 39
2.2　"水-蒸汽"给热系数实验 ……………………………………………… 39
　　2.2.1　实验目的 ……………………………………………………… 39
　　2.2.2　基本原理 ……………………………………………………… 39
　　2.2.3　装置与流程 …………………………………………………… 41
　　2.2.4　实验步骤 ……………………………………………………… 42
　　2.2.5　思考题 ………………………………………………………… 42

第三章　典型质量传递实验 …………………………………………………… 43
3.1　"乙醇-水"筛板精馏实验 ……………………………………………… 43
　　3.1.1　实验目的 ……………………………………………………… 43
　　3.1.2　实验原理 ……………………………………………………… 43
　　3.1.3　实验装置 ……………………………………………………… 46
　　3.1.4　实验仿真装置 …………………………………………………… 47
　　3.1.5　实验步骤 ……………………………………………………… 49
　　3.1.6　仿真实验步骤 …………………………………………………… 51
　　3.1.7　数据记录及处理 ………………………………………………… 55
　　3.1.8　仿真实验记录及处理 …………………………………………… 55
　　3.1.9　问题讨论 ……………………………………………………… 56
3.2　填料吸收塔 ……………………………………………………………… 56
　　3.2.1　实验目的 ……………………………………………………… 56
　　3.2.2　实验原理 ……………………………………………………… 56
　　3.2.3　实验装置 ……………………………………………………… 57

3.2.4 实验步骤	58
3.2.5 数据记录及处理	58
3.2.6 问题讨论	59
3.3 流化床干燥实验	59
3.3.1 实验目的	59
3.3.2 实验原理	60
3.3.3 实验装置	63
3.3.4 实验仿真装置	64
3.3.5 实验步骤	64
3.3.6 仿真实验步骤	65
3.3.7 数据记录及处理	66
3.3.8 问题讨论	66

Contents

Chapter 1 Typical Momentum Transfer Experiment ········· 68
  1.1   Determination of Characteristic Curve of Centrifugal Pump ········· 68
     1.1.1 Purpose of the Experiment ········· 68
     1.1.2 Principle of the Experiment ········· 68
     1.1.3 Experimental Devices ········· 70
     1.1.4 Simulation Devices of the Experiment ········· 70
     1.1.5 Experimental Procedure ········· 70
     1.1.6 Procedure of Simulation Experiment ········· 72
     1.1.7 Data Recording and Processing ········· 74
     1.1.8 Recording and Processing of Simulation Experiment ········· 75
     1.1.9 Discussion ········· 77
  1.2   Determination of Fluid Resistance in Pipe ········· 79
     1.2.1 Purpose of the Experiment ········· 79
     1.2.2 Principle of the Experiment ········· 79
     1.2.3 Experimental Devices ········· 81
     1.2.4 Simulation Devices of the Experiment ········· 83
     1.2.5 Experimental Procedure ········· 83
     1.2.6 Procedure of Simulation Experiment ········· 85
     1.2.7 Data Recording and Processing ········· 89
     1.2.8 Recording and Processing of Simulation Experiment ········· 89
     1.2.9 Discussion ········· 90
  1.3   Control Test of Flowmeter ········· 91
     1.3.1 Purpose of the Experiment ········· 91
     1.3.2 Principle of the Experiment ········· 91
     1.3.3 Experimental Devices ········· 94
     1.3.4 Simulation Devices of the Experiment ········· 95
     1.3.5 Experimental Procedure ········· 96
     1.3.6 Procedure of Simulation Experiment ········· 97
     1.3.7 Data Recording and Processing ········· 100

1.3.8	Recording and Processing of Simulation Experiment	100
1.3.9	Discussion	102

Chapter 2   Typical Heat Transfer Experiment ······ 103

2.1   "Air-water" Heat Transfer Experiment ······ 103

- 2.1.1   Purpose of the Experiment ······ 103
- 2.1.2   Principle of the Experiment ······ 103
- 2.1.3   Experimental Devices ······ 104
- 2.1.4   Experimental Procedure ······ 104
- 2.1.5   Data Recording and Outcome Processing ······ 105
- 2.1.6   Devices and Process of the Virtual Simulation Experiment ······ 106
- 2.1.7   Operation of Heat Transfer Virtual Simulation Experiment ······ 107
- 2.1.8   Data Processing of the Virtual Simulation Experiment ······ 110
- 2.1.9   Questions to Consider ······ 112

2.2   "Water-vapor" Heat Transfer Coefficient Experiment ······ 112

- 2.2.1   Purpose of the Experiment ······ 112
- 2.2.2   Basic Principle ······ 112
- 2.2.3   Device and Process ······ 114
- 2.2.4   Experimental Procedure ······ 115
- 2.2.5   Questions to Consider ······ 117

Chapter 3   Typical Mass Transfer Experiment ······ 118

3.1   "Ethanol-water" Sieve Plate Rectification Experiment ······ 118

- 3.1.1   Purpose of the Experiment ······ 118
- 3.1.2   Principle of the Experiment ······ 118
- 3.1.3   Experimental Devices ······ 122
- 3.1.4   Simulation Devices of the Experiment ······ 124
- 3.1.5   Experimental Procedure ······ 125
- 3.1.6   Procedure of Simulation Experiment ······ 128
- 3.1.7   Data Recording and Processing ······ 132
- 3.1.8   Recording and Processing of Simulation Experiment ······ 132
- 3.1.9   Discussion ······ 133

3.2   Filler Absorber ······ 134

- 3.2.1   Purpose of the Experiment ······ 134
- 3.2.2   Principle of the Experiment ······ 134
- 3.2.3   Experimental Devices ······ 135

    3.2.4   Experimental Procedure ··· 135
    3.2.5   Data Recording and Processing ··· 137
    3.2.6   Discussion ·· 137
3.3   Fluidized Bed Drying Experiment ·· 138
    3.3.1   Purpose of the Experiment ·· 138
    3.3.2   Principle of the Experiment ··· 138
    3.3.3   Experimental Devices ·· 142
    3.3.4   Simulation Devices of the Experiment ······································· 143
    3.3.5   Experimental Procedure ··· 143
    3.3.6   Procedure of Simulation Experiment ·· 144
    3.3.7   Data Recording and Processing ··· 145
    3.3.8   Discussion ·· 146

文献(**References**) ·· 147

第一章 典型动量传递实验

1.1 离心泵特性曲线的测定

1.1.1 实验目的

① 了解离心泵结构与特性，熟悉离心泵的操作方法；
② 掌握离心泵特性曲线测定方法；
③ 了解电动调节阀的工作原理和使用方法。

1.1.2 实验原理

离心泵的特性曲线是选择和使用离心泵的重要依据之一，其特性曲线是在恒定转速下泵的扬程 H、轴功率 N 及效率 η 与泵的流量 Q 之间的关系曲线，它是流体在泵内流动规律的宏观表现形式。由于泵内部流动情况复杂，不能用理论方法推导出泵的特性关系曲线，只能依靠实验测定。

(1) 扬程 H 的测定与计算

取离心泵进口真空表和出口压力表处两截面，列出伯努利方程：

$$z_1 + \frac{p_1}{\rho g} + \frac{u_1^2}{2g} + H = z_2 + \frac{p_2}{\rho g} + \frac{u_2^2}{2g} + \Sigma h_f \tag{1-1}$$

由于两截面间的管长较短，通常可忽略阻力项 Σh_f，速度平方差也很小，故可忽略，则有

$$H = (z_2 - z_1) + \frac{p_2 - p_1}{\rho g} = H_0 + H_1(表值) + H_2 \tag{1-2}$$

式中：$H_0 = z_2 - z_1$，表示泵出口和进口间的位差(m)；

其中，ρ 为流体密度(kg/m^3)；g 为重力加速度(m/s^2)；p_1、p_2 分别为泵进、出口的真空度和表压(Pa)；H_1、H_2 分别为泵进、出口的真空度和表压对应的压头(m)；u_1、u_2 分别为泵进、出口的流速(m/s)；z_1、z_2 分别为真空表、压力表的安装高度(m)。

由式(1-2)可知，只要直接读出真空表和压力表上的数值及两表的安装高度差，就可计算出泵的扬程。

(2) 轴功率 N 的测量与计算

$$N = N_{电} \cdot k \text{ (W)} \tag{1-3}$$

式中，$N_{电}$为电功率表显示值，k为电机传动效率，可取$k = 0.95$。

(3)效率η的计算

泵的效率η是泵的有效功率N_e与轴功率N的比值。有效功率N_e是单位时间内流体经过泵时所获得的实际功，轴功率N是单位时间内泵轴从电机得到的功，两者差异反映了水力损失、容积损失和机械损失的大小。泵的有效功率N_e可用式(1-4)计算：

$$N_e = HQ\rho g \tag{1-4}$$

故泵效率为：

$$\eta = \frac{HQ\rho g}{N} \times 100\% \tag{1-5}$$

(4)转速改变时的换算

泵的特性曲线是在一定转速下的实验测定所得。但是，实际上感应电动机在转矩改变时，其转速会有变化，这样随着流量Q的变化，多个实验点的转速n将有所差异，因此在绘制特性曲线之前，须将实测数据换算为某一定转速n'下(可取离心泵的额定转速为2900rpm)的数据。换算关系如下：

流量 $$Q' = Q \frac{n'}{n} \tag{1-6}$$

扬程 $$H' = H \left(\frac{n'}{n}\right)^2 \tag{1-7}$$

轴功率 $$N' = N \left(\frac{n'}{n}\right)^3 \tag{1-8}$$

效率 $$\eta' = \frac{Q'H'\rho g}{N'} = \frac{QH\rho g}{N} = \eta \tag{1-9}$$

1.1.3 实验装置

离心泵特性曲线测定装置流程如图1-1所示。

1.1.4 实验仿真装置

离心泵性能曲线仿真实验装置界面如图1-2所示。其中，泵的转速：2900r/min；额定扬程：20m；电机效率：93%；传动效率：100%；水温：25℃；泵进口管内径：41mm；泵出口管内径：35.78mm；两测压口之间的垂直距离：0.35m；涡轮流量计流量系数：75.78。

1.1.5 实验步骤

①清洗水箱，并加装实验用水。通过灌泵漏斗给离心泵灌水，排出泵内气体。

②检查各阀门开度和仪表自检情况，试开状态下检查电机和离心泵是否正常运转。开启离心泵之前先将出口阀关闭，当泵达到额定转速后方可逐步打开出口阀。

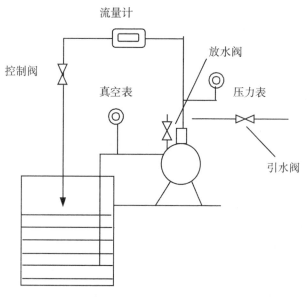

图 1-1 实验装置流程示意图

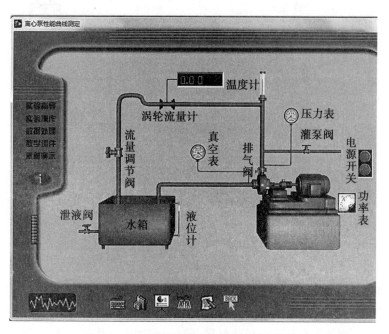

图 1-2 离心泵性能曲线仿真实验装置界面

③实验时,逐渐打开出口流量调节闸阀增大流量,待各仪表读数显示稳定后,读取相应数据。离心泵特性实验主要获取实验数据为:流量 Q、泵进口压力 p_1、泵出口压力

p_2、电机功率 $N_电$、泵转速 n 及流体温度 t 和两测压点间高度差 H_0($H_0=0.1\text{m}$)。

④改变出口流量调节闸阀的开度,测取10~15组数据后,可以停泵,同时记录下设备的相关数据(如离心泵型号、额定流量、额定转速、扬程和功率等),停泵前先将出口流量调节闸阀关闭。

1.1.6 仿真实验步骤

(1)灌泵

因为离心泵的安装高度在液面以上,所以在启动离心泵之前必须进行灌泵。如图1-3所示,打开灌泵阀。

在压力表上单击鼠标左键,即可放大读数(右键点击复原)。当读数大于0时,说明泵壳内已经充满水,但由于泵壳上部还留有一小部分气体,所以需要放气。调节排气阀开度大于0,即可放出气体,气体排尽后,会有液体涌出,如图1-3所示。此时关闭排气阀和灌泵阀,灌泵工作完成。

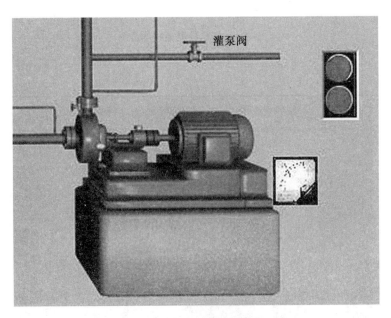

图1-3 离心泵灌泵示意图

(2)开泵

灌泵完成后,打开泵的电源开关,启动离心泵。注意:在启动离心泵时,主调节阀应关闭,如果主调节阀全开,会导致泵启动时功率过大,从而可能引发烧泵事故。

(3)建立流动

启动离心泵后,调节主调节阀的开度为100,如图1-4所示。

(4)读取数据

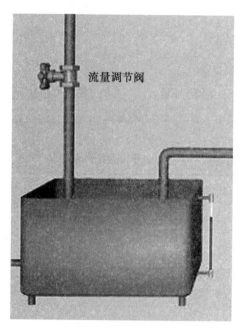

图 1-4　离心泵流量调节阀

等涡轮流量计的示数稳定后,即可读数。鼠标左键点击压力表、真空表和功率表,即可将其放大,以读取数据,如图 1-5 所示。

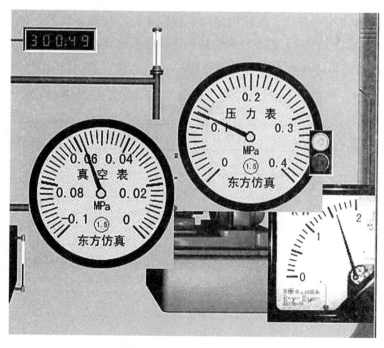

图 1-5　离心泵仪表示意图

1.1.7 数据记录及处理

① 记录实验原始数据见表1-1：

离心泵型号：_____；额定流量：_____；额定扬程：_____；额定功率：_____；泵进出口测压点高度差 H_0：_____；流体温度 t：_____

表1-1　　　　　　　　　　　　离心泵实验记录表

序号	流量 Q (m³/h)	泵进口压力 p_1(kPa)	泵出口压力 p_2(kPa)	电机功率 $N_电$(kW)	泵转速 n (r/min)

② 根据原理部分的公式，按比例定律校核转速后，计算各流量下的泵扬程、轴功率和效率，见表1-2。同时，需要分别绘制一定转速下的 H-Q、N-Q、η-Q 曲线；分析实验结果，判断泵最为适宜的工作范围。

表1-2　　　　　　　　　　　　数据处理结果

序号	流量 Q' (m³/h)	扬程 H' (m)	轴功率 N' (kW)	泵效率 η' (%)

1.1.8 仿真实验记录及处理

(1) 实验记录

鼠标左键点击实验主画面左边菜单中的"数据处理",可调出数据处理窗口,在原始数据页面按项目分别将数据填入记录表,也可在用点击"打印数据记录表"键所打印的数据记录表中记录数据,两者形式基本相同,注意单位换算。仿真实验记录表格如图1-6所示。注意:如果使用自动记录功能,则当用户点击"自动记录"键时,数据会被自动写入而不需手动填写。如果使用"自动记录"功能或已经将数据记录在数据库内,则可以跳过此步,如果是将数据记录在通过点击"打印数据记录表"键所打印的数据记录表内,则将数据填入表格中,如图1-7所示。

图1-6 离心泵仿真实验记录表格

(2) 数据处理

1) 数据计算

填好数据后,如果不采用"自动计算"功能,则可以在原始数据页找到计算所需的参数,如果要使用"自动计算"功能,在相应的计算结果页点击"自动计算"即可,数据即可自动计算并自动填入,如图1-8所示。

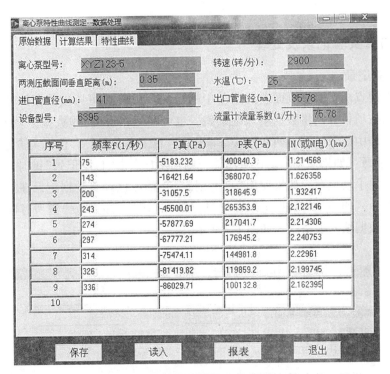

图 1-7 离心泵仿真实验数据表格

图 1-8 数据计算结果

2）特性曲线绘制

特性曲线结果如图 1-9 所示。计算完成后，如图 1-9 所示在曲线页点击"开始绘制"即可根据数据自动绘制出曲线。点击数据处理窗口下面一排按钮中的"打印"按钮，即可调出实验报表窗口（见图 1-10）。点击数据处理窗口下面一排按钮中的"保存"按钮，可保存原始数据到磁盘文件，并可点击"读入"按钮读入该数据文件。

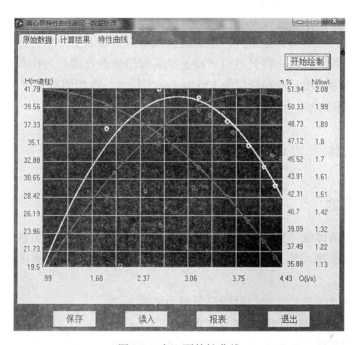

图 1-9　离心泵特性曲线

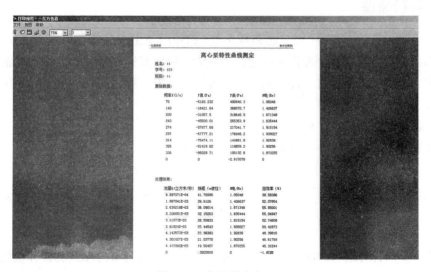

图 1-10　实验报表窗口

1.1.9 问题讨论

① 试从所测实验数据分析，离心泵在启动时为什么要关闭出口阀门？

② 启动离心泵之前为什么要引水灌泵？如果灌泵后依然启动不了，你认为可能的原因是什么？

③ 为什么用泵的出口阀门调节流量？这种方法有什么优缺点？是否还有其他方法调节流量？

④ 泵启动后，出口阀如果不开，压力表读数是否会逐渐上升？为什么？

⑤ 正常工作的离心泵，在其进口管路上安装阀门是否合理？为什么？

⑥ 试分析，用清水泵输送密度为 1200kg/m³ 的盐水，在相同流量下你认为泵的压力是否变化？轴功率是否变化？

1.2 管路流体阻力的测定

1.2.1 实验目的

① 掌握测定流体流经直管、管件和阀门时阻力损失的一般实验方法；

② 测定直管摩擦系数 λ 与雷诺准数 Re 的关系，验证在一般湍流区内 λ 与 Re 的关系曲线；

③ 测定流体流经管件、阀门时的局部阻力系数 ξ；

④ 学会倒 U 形压差计和涡轮流量计的使用方法；

⑤ 识别组成管路的各种管件、阀门，并了解其作用。

1.2.2 实验原理

流体通过由直管、管件(如三通和弯头等)和阀门等组成的管路系统时，由于黏性剪应力和涡流应力的存在，要损失一定的机械能。流体流经直管时所造成机械能损失称为直管阻力损失。流体通过管件、阀门时因流体运动方向和速度大小改变所引起的机械能损失称为局部阻力损失。

(1) 直管阻力摩擦系数 λ 的测定

流体在水平等径直管中稳定流动时，阻力损失为：

$$w_f = \frac{\Delta p_f}{\rho} = \frac{p_1 - p_2}{\rho} = \lambda \frac{l}{d} \frac{u^2}{2} \tag{1-10}$$

即

$$\lambda = \frac{2d\Delta p_f}{\rho l u^2} \tag{1-11}$$

式中，λ 为直管阻力摩擦系数，无因次；d 为直管内径(m)；Δp_f 为流体流经 l 米直管的压力降(Pa)；w_f 为单位质量流体流经 l 米直管的机械能损失(J/kg)；ρ 为流体密度

(kg/m^3);l 为直管长度(m);u 为流体在管内流动的平均流速(m/s)。

滞流(层流)时:

$$\lambda = \frac{64}{Re} \tag{1-12}$$

$$Re = \frac{du\rho}{\mu} \tag{1-13}$$

式中,Re 为雷诺准数,无因次;μ 为流体黏度(kg/(m·s))。

湍流时 λ 是雷诺准数 Re 和相对粗糙度(ε/d)的函数,须由实验确定。由式(1-11)可知,欲测定 λ,需确定 l、d,测定 Δp_f、u、ρ、μ 等参数。l、d 为装置参数(装置参数表格中给出),ρ、μ 通过测定流体温度,再查有关手册而得,u 通过测定流体流量,再由管径计算得到。采用涡轮流量计测流量 V(m^3/h)。

$$u = \frac{V}{900\pi d^2} \tag{1-14}$$

其中,Δp_f 可用 U 形管、倒置 U 形管、测压直管等液柱压差计测定,或采用差压变送器和二次仪表显示。当采用倒置 U 形管液柱压差计时,

$$\Delta p_f = \rho g R \tag{1-15}$$

式中,R 为水柱高度(m)。

当采用 U 形管液柱压差计时

$$\Delta p_f = (\rho_0 - \rho) g R \tag{1-16}$$

式中,R 为液柱高度(m);ρ_0 为指示液密度(kg/m^3)。

根据实验装置结构参数 l、d,指示液密度 ρ_0,流体温度 t_0(查流体物性 ρ、μ),及实验时测定的流量 V、液柱压差计的读数 R,通过式(1-14)、式(1-15)或式(1-16)、式(1-11)和式(1-13)求取 Re 和 λ,再将 Re 和 λ 标绘在双对数坐标图上。

(2)局部阻力系数 ξ 的测定

局部阻力损失通常有两种表示方法,即当量长度法和阻力系数法。

1)当量长度法

流体流过某管件或阀门时造成的机械能损失看作与某一长度为 l_e 的同直径的管道所产生的机械能损失相当,此折合的管道长度称为当量长度,用符号 l_e 表示。这样,就可以用直管阻力的公式来计算局部阻力损失,而且在管路计算时可将管路中的直管长度与管件、阀门的当量长度合并在一起计算,则流体在管路中流动时的总机械能损失 $\sum w_f$ 为:

$$\sum w_f = \lambda \frac{l + \sum l_e}{d} \frac{u^2}{2} \tag{1-17}$$

2)阻力系数法

流体通过某一管件或阀门时的机械能损失表示为流体在小管径内流动时平均动能的某一倍数,局部阻力的这种计算方法,称为阻力系数法。即

$$w'_f = \frac{\Delta p'_f}{\rho} = \xi \frac{u^2}{2} \tag{1-18}$$

故
$$\xi = \frac{2\Delta p'_f}{\rho u^2} \tag{1-19}$$

式中，ξ 为局部阻力系数，无因次；$\Delta p'_f$ 为局部阻力压强降(Pa)；(所测得的压降应扣除两测压口间直管段的压降，直管段的压降由直管阻力实验结果求取)。ρ 为流体密度(kg/m^3)；g 为重力加速度，$9.81m/s^2$；u 为流体在小截面管中的平均流速(m/s)。

待测的管件和阀门由现场指定。本实验采用阻力系数法表示管件或阀门的局部阻力损失。根据连接管件或阀门两端管径中小管的直径 d，指示液密度 ρ_0，流体温度 t_0(查流体物性 ρ、μ)及实验时测定的流量 V、液柱压差计的读数 R，通过式(1-14)、式(1-15)或式(1-16)、式(1-19)求取管件或阀门的局部阻力系数 ξ。

1.2.3 实验装置

(1) 实验装置

实验装置如图1-11所示。

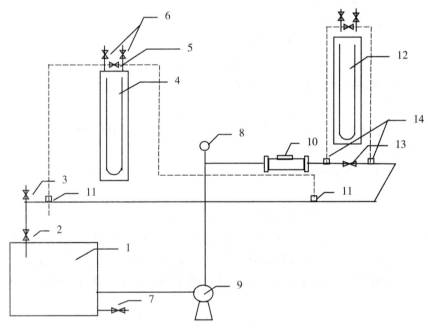

1：水箱；2：控制阀；3：放空阀；4：直管阻力测量U形压差计；5：平衡阀；
6：放空阀；7：排液阀；8：温度计；9：泵；10：涡轮流量计；11：直管段取压孔；
12：局部阻力测量U形压差计；13：闸阀；14：局部阻力取压孔

图1-11 流体流动阻力实验示意图

(2) 实验流程

实验对象部分是由贮水箱，离心泵，不同管径、材质的水管，各种阀门、管件，涡轮流量计和倒 U 形压差计等所组成的。管路部分有三段并联的长直管，分别用于测定局部阻力系数、光滑管直管阻力系数和粗糙管直管阻力系数。测定局部阻力部分使用不锈钢管，其上装有待测管件（闸阀）；光滑管直管阻力的测定同样使用内壁光滑的不锈钢管，而粗糙管直管阻力的测定对象为管道内壁较粗糙的镀锌管。流量使用涡轮流量计测量，将涡轮流量计的信号传给相应的显示仪表显示出转速，管路和管件的阻力采用倒 U 形差压计直接读出读数。

(3) 装置参数

装置参数见表 1-3。由于管子的材质存在批次的差异，所以可能会造成管径的不同，表 1-3 为管内径参数。

表 1-3　　　　　　　　　装 置 参 数

	名称	材质	管内径(mm)		测量段长度(cm)
			管路号	管内径	
装置	局部阻力	闸阀	1A	20.0	95
	光滑管	不锈钢管	1B	20.0	100
	粗糙管	镀锌铁管	1C	21.0	100

1.2.4　实验仿真装置

管路流体阻力的测定虚拟仿真软件界面如图 1-12 所示。其中，光滑管：玻璃管的内径=20mm，管长=1.5m，绝对粗糙度=0.002mm；粗糙管：镀锌铁管内径=20mm，管长=1.5m，绝对粗糙度=0.2mm；突然扩大管：细管内径=20mm，粗管内径=40mm；孔板流量计：开孔直径=12mm，孔流系数=0.62。

1.2.5　实验步骤

(1) 实验准备

① 清洗水箱，清除底部杂物，防止损坏泵的叶轮和涡轮流量计。关闭箱底侧排污阀，灌清水至离水箱上缘约 15 cm 高度，既可提供足够的实验用水又可防止出口管处水花飞溅。

② 接通控制柜电源，打开总开关电源及仪表电源，进行仪表自检。打开水箱与泵连接管路间的球阀，关闭泵的回流阀，全开转子流量计下的闸阀。如上步骤操作后，若泵吸不上水，可能是叶轮反转，首先检查有无缺相，一般可从指示灯判断三相电是否正

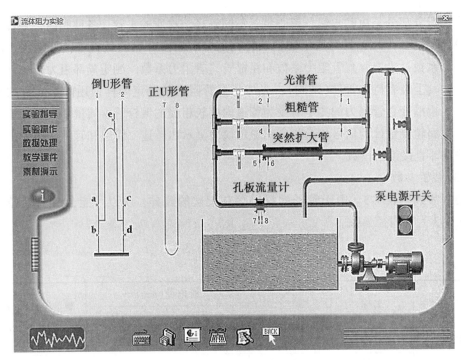

图 1-12 管路流体阻力的测定虚拟仿真软件界面

常。其次检查有无反相，需检查管道离心泵电机部分电源相序，调整三根火线中的任意两线插口即可。

（2）实验管路选择

选择实验管路，把对应的进口阀打开，并在出口阀最大开度下，保持全流量流动 5~10min。

（3）排气

先进行管路的引压操作。需打开实验管路均压环上的引压阀，对倒 U 形管进行操作如下，其结构如图 1-13 所示。

① 排出系统和导压管内的气泡。关闭管路总出口阀 5，使系统处于零流量、高扬程状态。关闭进气阀门 3 和平衡阀门 4。打开高压侧阀门 2 和低压侧阀门 1，使实验系统的水经过系统管路、导压管、高压侧阀门 2、倒 U 形管、低压侧阀门 1 排出系统。

② 玻璃管吸入空气。排净气泡后，关闭 1 和 2 两个阀门，打开平衡阀门 4 和出水活栓 5、进气阀门 3，使玻璃管内的水排净并吸入空气。

③ 平衡水位。关闭阀门 4、5、3，然后打开 1 和 2 两个阀门，让水进入玻璃管至平衡水位。此时系统中的出水阀门始终是关闭的，管路中的水在零流量时，U 形管内水位是平衡的，压差计即处于待用状态。

1.2 管路流体阻力的测定

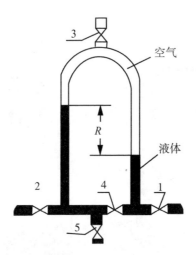

1：低压侧阀门；2：高压侧阀门；3：进气阀门；4：平衡阀门；5：出水活栓

图 1-13　倒 U 形管压差计

④ 被测对象在不同流量下对应的差压，就反映为倒 U 形管压差计的左右水柱之差。

(4) 流量调节

进行不同流量下的管路压差测定实验。让流量在 $0.8 \sim 4 m^3/h$ 范围内变化，建议每次实验变化 $0.5 m^3/h$ 左右。由小到大或由大到小调节管路总出口阀，每次改变流量，待流动达到稳定后，读取各项数据，共作 8~10 组实验点。主要获取实验参数：流量 Q、测量段压差 ΔP 及流体温度 t。

(5) 实验结束

实验完毕，关闭管路总出口阀，然后关闭泵开关和控制柜电源，将该管路的进口球阀和对应均压环上的引压阀关闭，清理装置(若长期不用，则管路残留水可从排空阀进行排空，水箱里的水也通过排水阀排空)。

1.2.6　仿真实验步骤

第一步：开泵，如图 1-14 所示。因为离心泵的安装高度比水的液面低，因此不需要灌泵。直接点击电源开关的绿色按钮接通电源，就可以启动离心泵，开始实验。

第二步：管道系统排气以及调节倒 U 形压差计，如图 1-15 所示。将管道中所有阀门都打开，使水在 3 个管路中流动一段时间，直到排净管道中的空气，然后点击倒 U 形压差计，软件画面会出现调节倒 U 形压差计的动画。最后关闭各阀门，开始试验操作。

第三步：测量光滑管数据，如图 1-16 所示。

(1) 光滑管建立流动

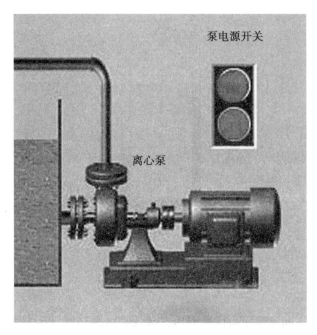

图 1-14 开泵示意图

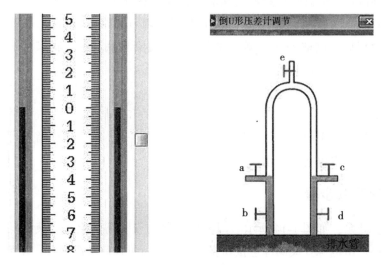

图 1-15 倒 U 形压差计

启动离心泵并调节完倒 U 形压差计后，如图 1-16 所示，依次调节阀 1、阀 2、阀 3 的开度大于 0，即可建立流动。关闭粗糙管和突然扩大管的球阀，打开光滑管的球阀，使水只在光滑管中流动。

（2）读取数据

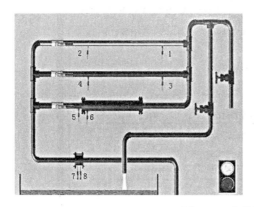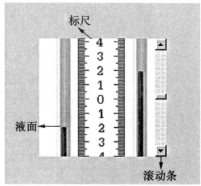

图 1-16　光滑管测试

鼠标左键点击正 U 形压差计或倒 U 形压差计，即可看到如图 1-16 所示的画面(红色液面只是作指示用，真实装置可能为其他颜色，如水银为银白色)。倒 U 形压差计的取压口与管道上的取压口相连，正 U 形压差计的取压口与孔板的取压口相连。用鼠标上下拖动滚动条即可读数。实验中每一管路均有一倒 U 形压差计，连续点击图中的倒 U 形压差计即可在 3 个倒 U 形压差计中切换。倒 U 形压差计上方的数字标出了与该管相连的管路。注意：读数为两液面高度差，单位为 mm。

(3) 记录数据

鼠标左键点击实验主画面左边菜单中的"数据处理"按钮，可调出数据处理窗口，如图 1-17 所示，点击原始数据页，按标准数据库操作方法在正 U 形压差计和倒 U 形压差计两栏中分别填入从正 U 形压差计和倒 U 形压差计所读取的数据。

注意：如果使用的是自动记录功能，则当点击"自动记录"按钮时，数据会被自动写入而不需手动填写。

(4) 记录多组数据

调节阀门开度以改变流量，重复上述第(2)、第(3)步，为了实验精度和回归曲线的需要至少应测量 10 组以上数据。完成后进入下一步——测量粗糙管数据。

第四步：测量粗糙管数据。

(1) 粗糙管建立流动

完成光滑管数据的测量和记录后，建立粗糙管的流动。

(2) 测量并记录数据

测量粗糙管的数据与测量光滑管的数据操作步骤相同，重复测量光滑管数据步骤的第(2)、第(3)、第(4)步，为了实验精度和回归曲线的需要至少应测量 10 组以上数据。完成后进入下一步——测量突然扩大管数据。

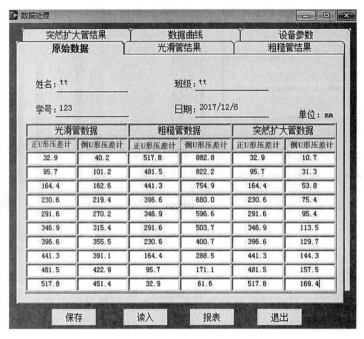

图 1-17　数据处理窗口

第五步：测量突然扩大管数据。

（1）突然扩大管建立流动

完成粗糙管数据的测量和记录后，建立突然扩大管的流动。

（2）突然扩大管数据的测量记录

测量突然扩大管的数据与测量光滑管的数据操作步骤相同，重复测量光滑管数据步骤的第（2）、第（3）、第（4）步，为了实验精度和回归曲线的需要至少应测量10组以上数据。完成后进入数据处理。

注意事项：

① 为了接近理想的光滑管，我们选用了玻璃管，实际上在普通实验室中很少采用玻璃管。

② 为了更好地回归处理数据，请尽量多地测量数据，并且尽量使数据分布在整个流量范围内。

③ 在层流范围内，用阀门按钮调节很难控制精度，请在阀门开度栏内自己输入开度数值（阀门开度小于5）。

④ 对于突然扩大管，我们做了简化，认为阻力系数是定值，不随 Re 变化。

1.2.7　数据记录及处理

将上述实验测得的数据填写到表1-4中：

直管基本参数：光滑管径_____；粗糙管径_____；局部阻力管径_____。

表1-4　　　　　　　　　　　实验数据记录表

序号	流量 (m³/h)	光滑管 mmH$_2$O			粗糙管 mmH$_2$O			局部阻力 mmH$_2$O		
		左	右	压差	左	右	压差	左	右	压差

1.2.8　仿真实验记录及处理

第一步：原始数据记录，如图1-18所示，注意：由于三组数据的格式相同，请注意不要混淆。

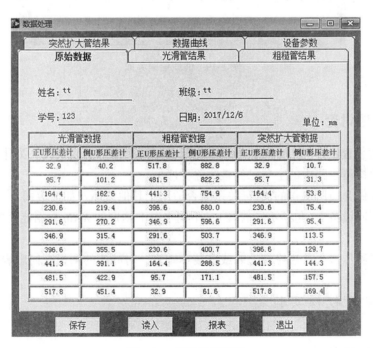

图1-18　原始数据记录

第二步：数据计算。填好数据后，如果不采用"自动计算"功能，则可以在数据处理的"设备参数"页得到计算所需的设备参数。如果要使用"自动计算"功能，在相应的计算结果页点击"自动计算"即可。数据即可自动计算并自动填入数据库中。

第三步：曲线绘制。计算完成后，在如图 1-19 所示的曲线页上点击"自动绘制"即可根据数据自动绘制出曲线。

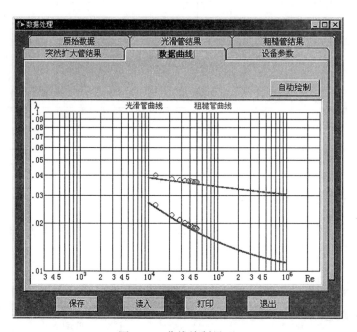

图 1-19　曲线绘制界面

1.2.9　问题讨论

① 在对装置做排气工作时，是否一定要关闭流程尾部的出口阀？为什么？
② 如何检测管路中的空气已经被排除干净？
③ 以水做介质所测得的 λ-Re 关系能否适用于其他流体？如何应用？
④ 在不同设备上（包括不同管径），不同水温下测定的 λ-Re 数据能否关联在同一条曲线上？
⑤ 如果测压口、孔边缘有毛刺或安装不垂直，对静压的测量有何影响？

1.3　流量计校核实验

1.3.1　实验目的

① 熟悉孔板流量计、文丘里流量计的构造、性能及安装方法。

② 掌握流量计容量法标定。
③ 测定孔板流量计、文丘里流量计的孔流系数与雷诺准数的关系。

1.3.2 实验原理

对非标准化的各种流量仪表在出厂前都必须进行流量标定，建立流量刻度标尺(如转子流量计)、给出孔流系数(如涡轮流量计)、给出校正曲线(如孔板流量计)。使用者在使用时，如工作介质、温度、压强等操作条件与原来标定时的条件不同，就需要根据现场情况，对流量计进行标定。孔板、文丘里流量计的收缩口面积都是固定的，而流体通过收缩口的压力降则随流量大小而变，据此来测量流量，因此，称其为变压头流量计。而另一类流量计中，当流体通过时，压力降不变，但收缩口面积却随流量而改变，故称这类流量计为变截面流量计，此类的典型代表是转子流量计。

1.3.2.1 孔板流量计

孔板流量计是应用最广泛的节流式流量计之一，本实验采用自制的孔板流量计测定液体流量，用容量法进行标定，同时测定孔流系数与雷诺准数的关系。孔板流量计是根据流体的动能和势能相互转化原理而设计的，流体通过锐孔时流速增加，使孔板前后产生压强差，可以通过引压管在压差计或差压变送器上显示。其基本构造如图1-20所示。

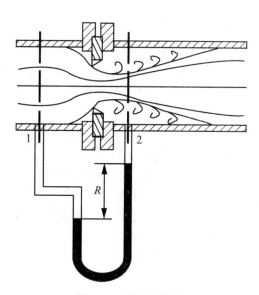

图1-20　孔板流量计

若管路直径为d_1，孔板锐孔直径为d_0，流体流经孔板前后所形成的缩脉直径为d_2，流体的密度为ρ，则根据伯努利方程，在界面1、2处有：

$$\frac{u_2^2 - u_1^2}{2} = \frac{p_1 - p_2}{\rho} = \frac{\Delta p}{\rho} \tag{1-20}$$

或
$$\sqrt{u_2^2 - u_1^2} = \sqrt{2\Delta p/\rho} \tag{1-21}$$

由于缩脉处位置随流速而变化，截面积 A_2 又难以知道，而孔板孔径的面积 A_0 是已知的，因此，用孔板孔径处流速 u_0 来替代式(1-21)中的 u_2，又考虑这种替代带来的误差以及实际流体局部阻力造成的能量损失，故需用系数 C 加以校正。式(1-21)可改写为：

$$\sqrt{u_2^2 - u_1^2} = C\sqrt{2\Delta p/\rho} \tag{1-22}$$

对于不可压缩流体，根据连续性方程可知 $u_1 = \frac{A_0}{A_1} u_0$，代入式(1-22)并整理可得：

$$u_0 = \frac{C\sqrt{2\Delta p/\rho}}{\sqrt{1 - \left(\frac{A_0}{A_1}\right)^2}} \tag{1-23}$$

令
$$C_0 = \frac{C}{\sqrt{1 - \left(\frac{A_0}{A_1}\right)^2}} \tag{1-24}$$

则式(1-23)可简化为：
$$u_0 = C_0\sqrt{2\Delta p/\rho} \tag{1-25}$$

根据 u_0 和 A_0 即可计算出流体的体积流量：

$$V = u_0 A_0 = C_0 A_0 \sqrt{2\Delta p/\rho} \tag{1-26}$$

或
$$V = u_0 A_0 = C_0 A_0 \sqrt{2gR(\rho_i - \rho)/\rho} \tag{1-27}$$

式中，V 为流体的体积流量，m^3/s；R 为 U 形压差计的读数，m；ρ_i 为压差计中指示液密度，kg/m^3；C_0 为孔流系数，无因次。

C_0 由孔板锐口的形状、测压口位置、孔径与管径之比和雷诺准数 Re 所决定，具体数值由实验测定。当孔径与管径之比为一定值时，Re 超过某个数值后，C_0 接近于常数。一般工业上定型的流量计，就是规定在 C_0 为定值的流动条件下使用。C_0 值范围一般为 0.6~0.7。孔板流量计安装时应在其上、下游各有一段直管段作为稳定段，上游长度至少应为 $10d_1$，下游为 $5d_2$。孔板流量计构造简单，制造和安装都很方便，其主要缺点是机械能损失大。由于机械能损失大，下游速度复原后，压力不能恢复到孔板前的值，称之为永久损失。d_0/d_1 的值越小，永久损失越大。

1.3.1.2 文丘里流量计

孔板流量计的主要缺点是机械能损失很大，为了克服这一缺点，可采用一渐缩渐括管，如图 1-21 所示，当流体流过这样的锥管时，不会出现边界层分离及漩涡，从而大大降低了机械能损失。这种管称为文丘里管。文丘里管收缩锥角通常取 15°~25°，扩大段锥角要取得小些，一般为 5°~7°，使流速改变平缓，因为机械能损失主要发生在管口突然扩大处。

1.3 流量计校核实验

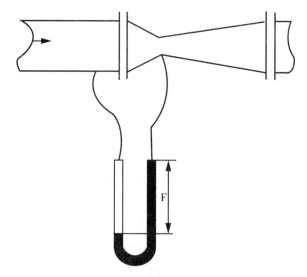

图 1-21　文丘里流量计

文丘里流量计测量原理与孔板流量计完全相同,只不过永久损失要小很多。流速、流量计算仍可用式(1-25)、式(1-26),式中 u_0 仍代表最小截面处(称为文氏喉)的流速。文丘里管的孔流系数 C_0 为 0.98~0.99。机械能损失约为：

$$w_f = 0.1u_0^2 \tag{1-28}$$

文丘里流量计的缺点是加工比孔板流量计复杂,因而造价高,且安装时需占去一定管长位置,但其永久损失小,故尤其适用于低压气体的输送。

1.3.3 实验装置

实验装置如图 1-22 所示。主要部分由循环水泵、流量计、U 形压差计、温度计和水槽等组成,实验主管路为 1 寸不锈钢管(内径 25mm)。

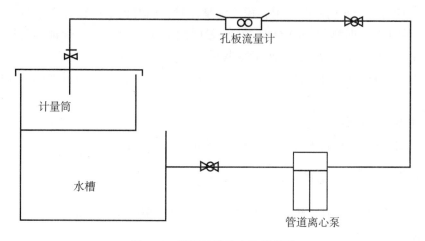

图 1-22　流量计校核实验示意图

1.3.4 实验仿真装置

流量计校核虚拟仿真实验界面如图 1-23 所示,设备参数:计量桶面积为 1m^2;管道直径为 30mm;孔板开孔直径为 20mm。

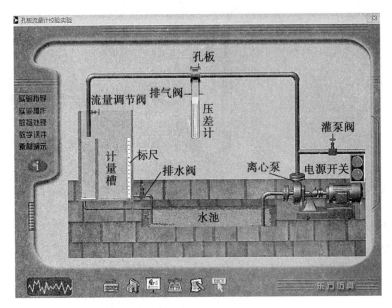

图 1-23 流量计校核虚拟仿真实验界面

1.3.5 实验步骤

① 熟悉实验装置,了解各阀门的位置及作用。

② 对装置中有关管道、导压管、压差计进行排气,使倒 U 形压差计处于工作状态。

③ 对应每一个阀门开度,用容积法测量流量,同时记下压差计的读数,按由小到大的顺序在小流量时测量 8~9 个点,大流量时测量 5~6 个点。为保证标定精度,最好再从大流量到小流量重复一次,然后取其平均值。

④ 测量流量时应保证每次测量中,计量桶液位差不小于 100mm 或测量时间不少于 40 s。

⑤ 主要计算过程如下:

a. 根据体积法(秒表配合计量筒)算得流量 $V(\text{m}^3/\text{h})$;

b. 根据 $u = \dfrac{4V}{\pi d^2}$,求得 u 值;

c. 读取流量 V(由闸阀开度调节)对应下的压差计高度差 R,根据 $u_0 = C_0\sqrt{2\Delta p/\rho}$ 和 $\Delta p = \rho g R$,求得 C_0 值。

d. 根据 $Re = \dfrac{du\rho}{\mu}$，求得雷诺准数，其中 d 取对应的 d_0 值。

e. 在坐标纸上分别绘出孔板流量计和文丘里流量计的 $C_0\text{-}Re$ 图。

1.3.6 仿真实验步骤

第一步：灌泵，如图 1-24 所示。因为离心泵的安装高度在液面以上，所以在启动离心泵之前必须进行灌泵。因为本实验的重点在流量计，而不是离心泵，所以对灌泵进行了简化，如图 1-24 所示，只要调节灌泵阀开度大于 0，等待 10 秒以上，然后关闭，系统就会认为已经完成了灌泵操作。

图 1-24　灌泵

第二步：开泵，如图 1-25 所示。灌泵工作完成后，点击电源开关的绿色按钮接通电源，就可以启动离心泵，并开始工作。

图 1-25　开泵

第三步：启动离心泵后，调节主调节阀的开度为100，即可建立流动，如图1-26所示。

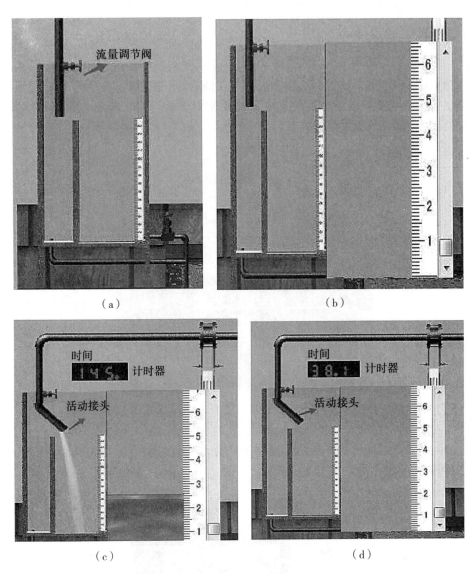

图 1-26　流量调节

第四步：读取数据。用鼠标左键点击标尺，即可调出标尺的读数画面，先记录下液面的初始高度。鼠标右键点击可关闭标尺画面。然后用鼠标左键点击活动接头，即可把水流引向计量槽，可以看到液面开始上升，同时计时器会自动开始计时。当液面上升到一定高度时，鼠标左键点击活动接头，将其转到泄液部分，同时计时器也会自动停止。此时记录下液面高度和计时器读数。用鼠标左键点击压差计，用鼠标拖动滚动条，读取压差。

第五步：记录数据，如图 1-27 所示。

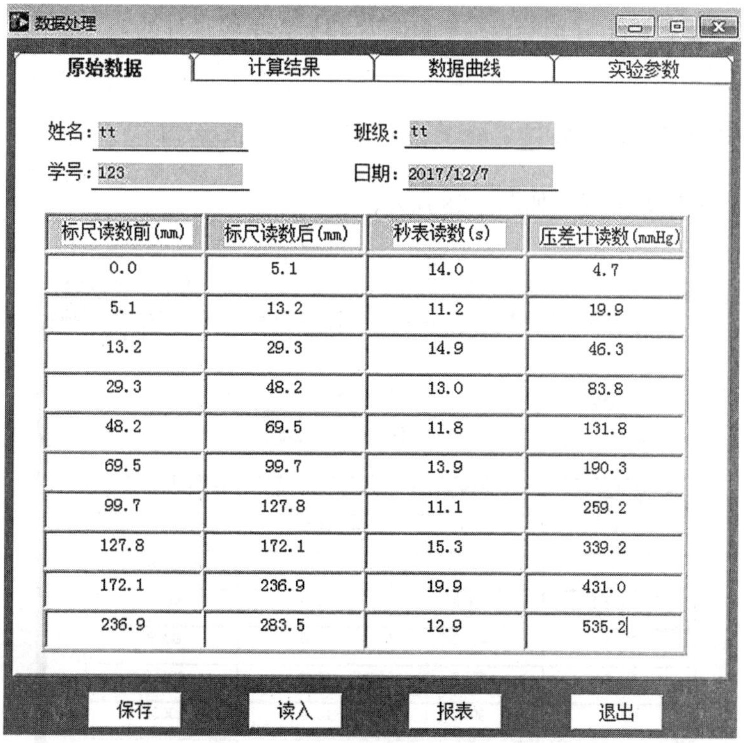
数据处理
原始数据
计算结果
数据曲线
实验参数
姓名：tt
班级：tt
学号：123
日期：2017/12/7
标尺读数前(mm) 标尺读数后(mm) 秒表读数(s) 压差计读数(mmHg)
0.0 5.1 14.0 4.7
5.1 13.2 11.2 19.9
13.2 29.3 14.9 46.3
29.3 48.2 13.0 83.8
48.2 69.5 11.8 131.8
69.5 99.7 13.9 190.3
99.7 127.8 11.1 259.2
127.8 172.1 15.3 339.2
172.1 236.9 19.9 431.0
236.9 283.5 12.9 535.2
保存
读入
报表
退出

图 1-27　数据记录表

鼠标左键点击实验主画面左边菜单中的“数据处理”，可调出数据处理窗口，点击原始数据页，按标准数据库操作方法在正 U 形压差计和倒 U 形压差计两栏中分别填入从正 U 形压差计和倒 U 形压差计所读取的数据，也可在用点击“打印数据记录表”按钮所打印的数据记录表中记录数据。

注意：如果使用自动记录功能，则当用户点击“自动记录”键时，数据会被自动写入而不需手动填写。为了更好地表现孔流系数 C_0在 Re 比较小时随 Re 的变化，我们把实验中的流量定得很低，以获得较小的 Re。另外，一般流量计校验实验是在孔流系数几乎不变的范围内测定多次取平均值，以得到 C_0，而不采用 C_0随 Re 的变化关系。因此，如果用手动记录数据和计算，就会出现很大的误差，用自动计算可以得到比较好的结果。

第六步：记录多组数据。调节主调节阀的开度以改变流量，然后重复上述第 4~5 步，为了实验精度和回归曲线的需要，至少要测 10 组数据。记录完毕后进入数据处理。

1.3.7 数据记录及处理

① 将所有原始数据及计算结果列成表格，并附上计算示例。
② 在单对数坐标纸上分别绘出孔板流量计和文丘里流量计的 C_0-Re 图。
③ 讨论实验结果。

1.3.8 仿真实验记录及处理

第一步：记录原始数据，如图 1-28 所示。

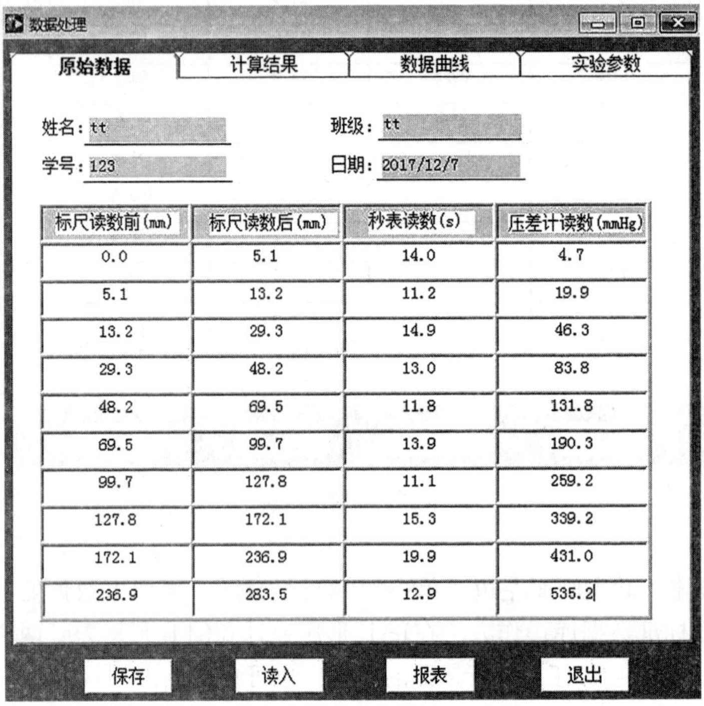

图 1-28 原始数据记录

如果使用"自动记录"功能或已经将数据记录在数据库内，则可以跳过此步，如果用户是将数据记录在用点击"打印数据记录表"键所打印的数据记录表内，那就需要参阅数据记录将所有数据填入数据库中。

第二步：数据计算，如图 1-29 所示。

如果要使用"自动计算"功能，在相应的计算结果页点击"自动计算"即可，如图1-29 所示。数据即可自动计算并自动填入数据库。

第三步：曲线绘制。计算完成后，如图 1-30 所示，在曲线页点击"自动绘制"按钮即可根据数据自动绘制出曲线。

1.3 流量计校核实验

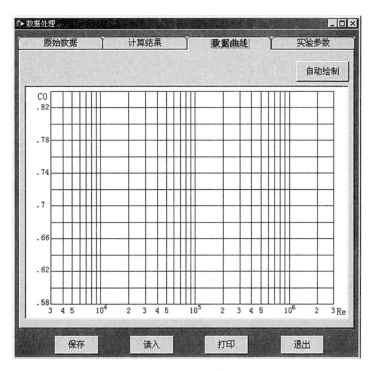

图 1-29　自动计算结果

图 1-30　数据曲线

1.3.9 问题讨论

① 孔流系数与哪些因素有关？
② 孔板、文丘里流量计安装时各应注意什么问题？
③ 如何检查系统排气是否完全？
④ 从实验中，可以直接得到 ΔR-V 的校正曲线，经整理后也可以得到 C_0-Re 的曲线，这两种表示方法各有什么优点？

第二章　典型热量传递实验

2.1 "空气-水"气液传热实验

2.1.1 实验目的

① 掌握传热系数 K 的测定方法。
② 学会换热器的操作方法。

2.1.2 实验原理

换热器在工业生产中是经常使用的换热设备。热流体借助于传热壁面，将热量传递给冷流体，以满足生产工艺的要求。影响换热器传热量的参数有传热面积、平均温度差和传热系数三要素。为了合理选用或设计换热器，应对其性能有充分的了解。除了查阅文献外，换热器性能实测是重要的途径之一。传热系数是度量换热器性能的重要指标。为了提高能量的利用率，提高换热器的传热系数以强化传热过程，在生产实践中是经常遇到的问题。管换热器是一种间壁式的传热装置，冷热液体间的传热过程由热流体对壁面的对流传热、间壁的固体热传导和壁面对冷流体的对流传热三个传热子过程组成，如图2-1所示。

传热速率由式(2-1)计算：

$$Q = KA \frac{(T_{出} - t_{进}) - (T_{进} - t_{出})}{\ln \dfrac{T_{出} - t_{进}}{T_{进} - t_{出}}} = c_{pg} q_{vg} \rho (T_{进} - T_{出}) \tag{2-1}$$

式中，K 为传热系数（$W/m^2 \cdot ℃$）；A 为传热面积（m^2）；$T_{进}$ 为热流体进口温度（℃）；$T_{出}$ 为热流体出口温度（℃）；$t_{进}$ 为冷流体进口温度（℃）；$t_{出}$ 为冷流体出口温度（℃）；c_{pg} 为热流体定压热容（$J/kg \cdot ℃$）；q_{vg} 为冷流体体积流量（m^3/s）；ρ 为热流体密度（kg/m^3）。

2.1.3 实验装置

本实验冷流体是水，热流体是空气。冷流体自冷流体源来，经转子流量计测量流量，温度计测量进口温度后，进入换热器壳程，换热后在出口处测量其出口温度。热流

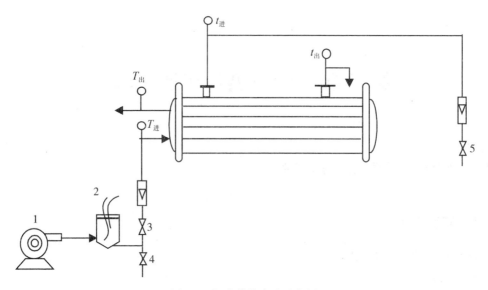

图 2-1 气液传热实验示意图

体自风源来,经转子流量计测量流量后,进入加热到 120℃流入换热器的管程,并在入口处测量其进口温度,在出口处测量其出口温度。实验装置如图 2-2 所示。

图 2-2 换热实验装置图

2.1.4 实验步骤

① 开通冷流体源，利用调节阀 5 调节冷流体流量。

② 开通风源 1，打开调节阀 4，由调节阀 3 调节空气流量，接通电源，在智能温度调节仪表 AI-708 上设定控制温度为 100~120℃。

③ 维持冷热流体流量不变，热空气进口温度在一定时间内（约 10 分钟）基本不变时，可记取有关数据。

④ 测定传热系数 K 时，在维持冷流体流量不变的情况下，根据实验步骤要求，改变热空气流量若干次。

⑤ 实验结束，关闭加热电源，待热空气温度降至 50℃ 以下，关闭冷热流体调节阀，并关闭冷热流体源。

2.1.5 数据记录与结果处理

请将实验数据记录在表 2-1 中。传热面积 A _____ m^2；冷流体体积流量 _____ L/h。

表 2-1　　　　　　　　　　　实验记录表

参数 序号	$T_{进}$(℃)	$T_{出}$(℃)	$t_{进}$(℃)	$t_{出}$(℃)	q_{vg}(m³/s)	K(W/(m²·℃))
1						
2						
3						
4						
5						
6						
7						

2.1.6 虚拟仿真实验设备流程

本装置流程如图 2-3 所示，冷水经由泵、U 形压差计，进入换热器内管，并与套管环隙中蒸汽换热。冷水流量可用流量控制阀调节。蒸汽由蒸汽发生器上升进入套管环隙，与内管中冷水换热。放气阀门用于排放不凝性气体。在铜管之前设有一定长度的稳定段，是为消除端效应。铜管两端用塑料管与管路相连，是为消除热应力。本实验装置冷水走内管，蒸汽走环隙（玻璃管）。水的进、出口温度和管壁温度分别由铂电阻测得。测量水进、出口温度的铂电阻应置于进、出口的中心。测量管壁温度的铂电阻用导热绝缘胶固定在内管外壁两端。孔板流量计的压差由 U 形压差计测得。本实验蒸汽发生器由不锈钢制成,并安有玻璃液位计。发生器的热功率为1.5kW。

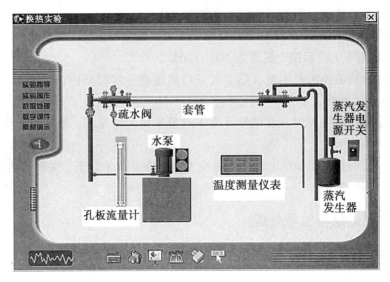

图 2-3　换热虚拟仿真实验界面

2.1.7　换热虚拟仿真实验操作

第一步:点击电源开关的绿色按钮,启动水泵,水泵为换热器的管程提供水源,如图 2-4 所示。

图 2-4　电源启动位置

第二步:打开进水阀,开泵后,调节进水阀至微开,这时换热器的管程中就有水流动了,如图 2-5 所示。

图 2-5 进水阀门开关

第三步:打开蒸汽发生器,蒸汽发生器的开关在蒸汽发生器的右侧。鼠标左键单击开关,这时蒸汽发生器就通电开始加热,并向换热器的壳程供汽,如图 2-6 所示。

图 2-6 蒸汽发生器

第四步:打开放汽阀,排出残余的不凝气体,使在换热器壳程中的蒸汽流动通畅,如图 2-7 所示。

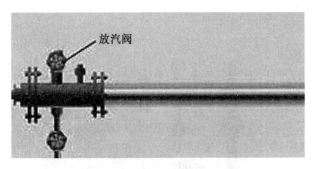

图 2-7 放汽阀

第五步：读取水的流量。在图中点击孔板流量计的压差计，出现读数画面。读取压差计读数，经过计算可得冷水的流量，如图 2-8 所示。

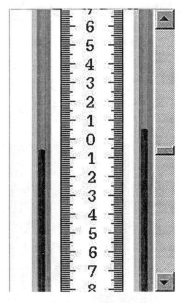

图 2-8 水流量

第六步：读取温度。在换热管或者测温仪上点击会出现温度读数画面。读取各处温度数值。其中，温度节点 1~9 的温度为观察温度分布用，在数据处理中用不到。蒸汽进出口及水进出口的温度需要记录。按"自动记录"按钮可由计算机自动记录实验数据。按"退出"按钮关闭温度读取画面，如图 2-9 所示。

第七步：记录多组数据，改变进水阀开度，重复以上步骤，读取 8~10 组数据。实验结束后，先关闭蒸汽发生器，再关进水阀。

2.1 "空气-水"气液传热实验

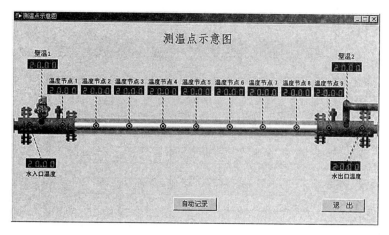

图 2-9 温度读取

2.1.8 虚拟仿真数据处理

第一步：原始数据记录。原始数据页面如图 2-10 所示，通过该页面能在数据处理中输入、编辑原始数据。

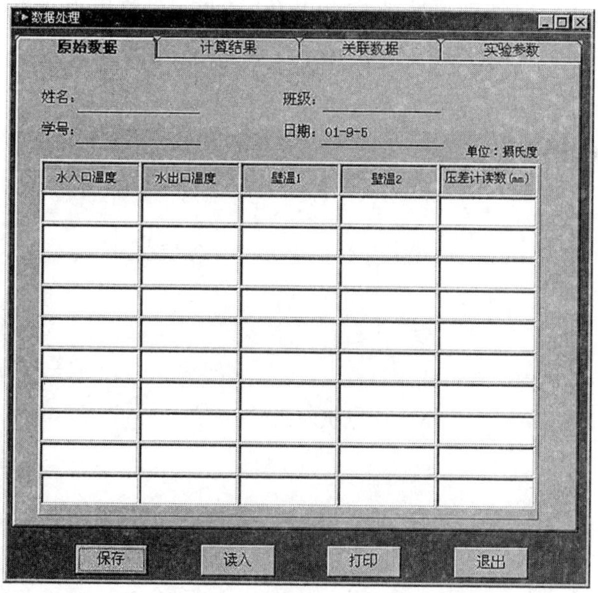

图 2-10 原始数据记录

第二步：数据计算。如果要使用自动计算功能，在相应的计算结果页点击"自动计算"按钮即可，如图 2-10 所示。数据即可自动计算并自动填入数据库。使用手动计算需要的设备参数可参见设备参数页，如图 2-11 所示。

第二章 典型热量传递实验

图 2-11 计算结果表

第三步：关联式。自动计算完成后，可在"关联式"菜单点击"自动关联"按钮，自动给出准数关联式(即给出图 2-12 中的 0.000 及 0.00 处的数值)。

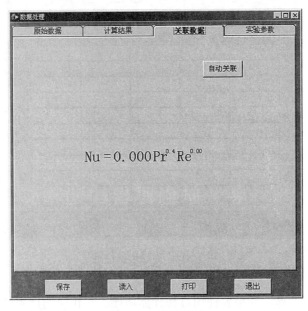

图 2-12 关联数据

2.1.9 思考题

当冷流体流量不变，改变热流体流量，试分析传热速率 Q 及传热系数 K 如何变化。

2.2 "水-蒸汽"给热系数实验

2.2.1 实验目的

① 了解间壁式传热元件，掌握给热系数测定的实验方法；
② 观察水蒸气在水平管外壁上的冷凝现象，测定水在圆形直管内的强制对流给热系数；
③ 了解影响给热系数的因素和强化传热的途径；
④ 了解热电阻测温方法、涡轮流量计测流量的方法，学会使用变频器。

2.2.2 基本原理

在工业生产过程中，大多数情况下，冷、热流体系通过固体壁面(传热元件)进行热量交换，称为间壁式换热。如图 2-13 所示，间壁式传热过程由热流体对固体壁面的对流传热、固体壁面的热传导和固体壁面对冷流体的对流传热组成。

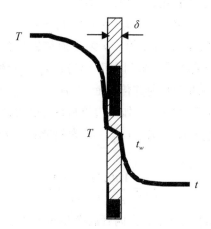

图 2-13 间壁式传热过程示意图

间壁式传热元件在传热过程达到稳态后，有

$$Q = m_1 c_{p1}(T_1 - T_2) = m_2 c_{p2}(t_2 - t_1)$$
$$= \alpha_1 A_1 (T - T_W)_M = \alpha_2 A_2 (t_W - t)_m \tag{2-2}$$

式中，Q 为传热量(J/s)；m_1 为热流体的质量流率(kg/s)；c_{p1} 为热流体的比热(J/(kg·℃))；T_1 为热流体的进口温度(℃)；T_2 为热流体的出口温度(℃)；m_2 为冷流体的质量流率(kg/s);c_{p2}为冷流体的比热(J/(kg·℃));t_1为冷流体的进口温度(℃);t_2为冷流体的出口温度(℃);α_1为热流体与固体壁面的对流传热系数(W/(m²·℃));A_1为热流体侧的对流传热面积(m²);$(T-T_W)_m$为热流体与固体壁面的对数平均温差(℃);α_2为冷流体与固体壁面的对流传热系数(W/(m²·℃));A_2为冷流体侧的对流传热面积(m²);$(t_W-t)_m$为固体壁面与冷流体的对数平均温差(℃)。

热流体与固体壁面的对数平均温差可由式(2-3)计算:

$$(T-T_W)_m = \frac{(T_1-T_{W1})-(T_2-T_{W2})}{\ln\dfrac{T_1-T_{W1}}{T_2-T_{W2}}} \qquad (2-3)$$

式中,T_{W1}为热流体进口处热流体侧的壁面温度(℃);T_{W2}为热流体出口处热流体侧的壁面温度(℃)。

固体壁面与冷流体的对数平均温差可由式(2-4)计算:

$$(t_W-t)_m = \frac{(t_{W1}-t_1)-(t_{W2}-t_2)}{\ln\dfrac{t_{W1}-t_1}{t_{W2}-t_2}} \qquad (2-4)$$

式中,t_{W1}为冷流体进口处冷流体侧的壁面温度(℃);t_{W2}为冷流体出口处冷流体侧的壁面温度(℃)。

在套管换热器中,换热桶内通水蒸气,内铜管管内通水,水蒸气在铜管表面冷凝放热而加热水,在传热过程达到稳定后,见式(2-5):

$$V\rho C_P(t_2-t_1) = \alpha_2 A_2(t_W-t)_m \qquad (2-5)$$

式中,V为冷流体体积流量(m³/s);ρ为冷流体密度(kg/m³);C_P为冷流体比热(J/(kg·℃));t_1、t_2为冷流体进、出口温度(℃);α_2为冷流体对内管内壁的对流给热系数(W/(m²·℃));A_2为内管的内壁传热面积(m²);$(t_W-t)_m$为内壁与流体间的对数平均温度差,参照式(2-3)可得(℃)。如果内管材料导热性能很好,即λ值很大,且管壁厚度很薄时,可认为$T_{W1}=t_{W1}$,$T_{W2}=t_{W2}$,即为所测得的该点的壁温,则由式(2-1)可得:

$$\alpha_2 = \frac{V\rho C_P(t_2-t_1)}{A_2(t_W-t)_m} \qquad (2-6)$$

若能测得被加热流体的V、t_1、t_2,内管的换热面积A_2,壁温t_{W1}、t_{W2},则可通过式(2-6)算得实测的冷流体在管内的对流给热系数α_2。

对于流体在圆形直管内作强制湍流对流传热时,传热准数经验式为:

$$N_u = 0.023 Re^{0.8} P_r^n \qquad (2-7)$$

式中,N_u为努塞尔数,$N_u = \dfrac{\alpha d}{\lambda}$,无因次;$Re$为雷诺准数,$Re = \dfrac{du\rho}{\mu}$,无因次;$P_r$为普兰特数,$P_r = \dfrac{c_p\mu}{\lambda}$,无因次;式(2-7)的适用范围为:$Re = 1.0\times10^4 \sim 1.2\times10^5$,

$P_r = 0.7 \sim 120$，管长与管内径之比 $\frac{L}{d} \geq 60$。当流体被加热时，$n = 0.4$，流体被冷却时，$n = 0.3$。α 为流体与固体壁面的对流传热系数（W/(m²·℃)）；d 为换热管内径（m）；λ 为流体的导热系数（W/(m·℃)）；u 为流体在管内流动的平均速度（m/s）；ρ 为流体的密度（kg/m³）；μ 为流体的黏度（Pa·s）；c_p 为流体的比热（J/(kg·℃)）。故可由实验获取的数据点拟合出相关准数后，即可作出曲线，并与经验公式的曲线对比以验证实验效果。

2.2.3 装置与流程

本实验装置由蒸汽发生器、玻璃转子流量计、套管换热器及温度传感器、温度显示仪表等构成，如图 2-14 所示。装置参数：紫铜管规格 12×2mm，即内径为 8mm，长度为 1m；冷流体最大流量为 400 L/h，涡轮流量计的测量下限为 40 L/h。"水-蒸汽"体系：来自蒸汽发生器的水蒸气进入套管换热器，与来自水箱的水进行热交换，冷凝水经管道排入地沟。冷水经增压泵和转子流量计进入套管换热器内管（紫铜管），水流量可用调节阀调节，热交换后进入下水道。注意，本实验中冷流体并非循环使用，实验过程中需要给水箱连续通水源。

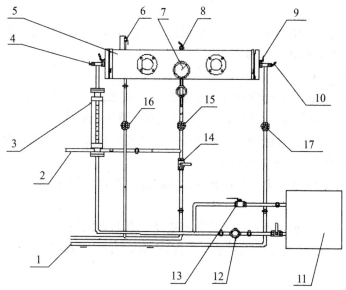

1：进水管路；2：蒸汽入口；3：玻璃转子流量计；4：冷流体进口温度；5：套管换热器；
6：惰性气体；7：蒸汽进口压力表；8：蒸汽温度；9：蒸汽出口温度；10：冷流体出口温度；
11：水箱；12：水泵；13：旁路球阀；14，16：冷凝水排空阀；15：蒸汽进口调节球阀；
17：流体流量调节阀

图 2-14 "水-蒸汽"给热系数实验装置

2.2.4 实验步骤

(1) 手动操作

① 检查仪表、水泵、蒸汽发生器及测温点是否正常，检查进系统的蒸汽调节阀是否关闭。

② 打开总电源开关、仪表电源开关(由教师启动蒸汽发生器和打开蒸汽总阀)。

③ 启动水泵。

④ 调节手动调节阀的开度，阀门全开使水流量达到最大。

⑤ 排除蒸汽管线中原积存的冷凝水(方法是：关闭进系统的蒸汽调节阀，打开蒸汽管冷凝水排放阀)。

⑥ 排净后，关闭蒸汽管冷凝水排放阀，打开进系统的蒸汽调节阀，使蒸汽缓缓进入换热器环隙以加热套管换热器，再打开换热器冷凝水排放阀(冷凝水排放阀不要开启过大，以免蒸汽泄漏)，使环隙中冷凝水不断地排至地沟。

⑦ 仔细调节进系统蒸汽调节阀的开度，使蒸汽压力稳定保持在 0.1MPa 左右(可通过微调惰性气体排空阀使压力达到需要的值)，以保证在恒压条件下操作，再根据测试要求，由大到小逐渐调节水流量手动调节阀的开度，合理确定 6 个实验点。待流量和热交换稳定后，分别读取冷流体流量、冷流体进出口温度、冷流体进出口壁温以及蒸汽温度。

⑧ 最后，首先关闭蒸汽调节阀，切断设备的蒸汽来路，关闭蒸汽发生器(由教师完成)、仪表电源开关及切断总电源。

(2) 注意事项

① 一定要在套管换热器内管输入一定量的空气或水，方可开启蒸汽阀门，且必须在排除蒸汽管线上原先积存的凝结水后，方可把蒸汽通入套管换热器中。

② 开始通入蒸汽时，要缓慢打开蒸汽阀门，使蒸汽徐徐流入换热器中，逐渐加热，由"冷态"转变为"热态"不得少于 10min。

③ 操作过程中，蒸汽压力一般控制在 0.3MPa(表压)以下，因为在此条件下压力比较容易控制。

④ 测定各参数时，必须是在稳定传热状态下，并且随时注意惰性气体的排空和压力表读数的调整。一般热稳定时间都至少需保证 5min 以上，以保证数据的可靠性。

2.2.5 思考题

① 实验中冷流体和蒸汽的流向对传热效果有何影响？

② 在计算冷流体质量流量时所用到的密度值与求雷诺准数时的密度值是否一致？它们分别表示什么位置的密度，应在什么条件下进行计算？

③ 实验过程中，冷凝水不及时排走，会产生什么影响？如何及时排走冷凝水？如果采用不同压强的蒸汽进行实验，对 α 关联式有何影响？

第三章 典型质量传递实验

3.1 "乙醇-水"筛板精馏实验

3.1.1 实验目的

① 了解筛板精馏塔及其附属设备的基本结构,掌握精馏过程的基本操作方法。
② 学会判断系统达到稳定的方法,掌握测定塔顶、塔釜溶液浓度的实验方法。
③ 学习测定精馏塔全塔效率和单板效率的实验方法,研究回流比对精馏塔分离效率的影响。

3.1.2 实验原理

(1) 全塔效率 E_T

全塔效率又称总板效率,是指达到指定分离效果所需理论板数与实际板数的比值,即

$$E_T = \frac{N_T - 1}{N_P} \tag{3-1}$$

式中,N_T 为完成一定分离任务所需的理论塔板数,包括蒸馏釜;N_P 为完成一定分离任务所需的实际塔板数,例如 $N_P = 10$。

全塔效率简单地反映了整个塔内塔板的平均效率,说明了塔板结构、物性系数、操作状况对塔分离能力的影响。对于塔内所需理论塔板数 N_T,可由已知的双组分物系平衡关系,以及实验中测得的塔顶、塔釜出液的组成,回流比 R 和热状况 q 等,用图解法求得。

(2) 单板效率 E_M

单板效率又称莫弗里板效率,是指气相或液相经过一层实际塔板前后的组成变化值与经过一层理论塔 x_{n-1} 前后的组成变化值之比。按气相组成变化表示的单板效率为:

$$E_{MV} = \frac{y_n - y_{n+1}}{y_n^* - y_{n+1}} \tag{3-2}$$

按液相组成变化表示的单板效率为:

$$E_{ML} = \frac{x_{n-1} - x_n}{x_{n-1} - x_n^*} \quad (3\text{-}3)$$

式中，y_n、y_{n+1} 分别为离开第 n、$n+1$ 块塔板的气相组成，摩尔分数；x_{n-1}、x_n 分别为离开第 $n-1$、n 块塔板的液相组成，摩尔分数；y_n^* 为与 x_n 成平衡的气相组成，摩尔分数；x_n^* 为与 y_n 成平衡的液相组成，摩尔分数。

（3）图解法求理论塔板数 N_T

图解法又称麦卡勃-蒂列（McCabe-Thiele）法，简称 M-T 法，其原理与逐板计算法完全相同，只是将逐板计算过程在 x–y 图上直观地表示出来。

精馏段的操作线方程为：

$$y_{n+1} = \frac{R}{R+1} x_n + \frac{x_D}{R+1} \quad (3\text{-}4)$$

式中，y_{n+1} 为精馏段第 $n+1$ 块塔板上升的蒸汽组成，摩尔分数；x_n 为精馏段第 n 块塔板下流的液体组成，摩尔分数；x_D 为塔顶馏出液的液体组成，摩尔分数；R 为泡点回流下的回流比。

提馏段的操作线方程为：

$$y_{m+1} = \frac{L'}{L' - W} x_m - \frac{W x_W}{L' - W} \quad (3\text{-}5)$$

式中，y_{m+1} 为提馏段第 $m+1$ 块塔板上升的蒸汽组成，摩尔分数；x_m 为提馏段第 m 块塔板下流的液体组成，摩尔分数；x_W 为塔底釜液的液体组成，摩尔分数；L' 为提馏段内下流的液体量(kmol/s)；W：釜液流量(kmol/s)。

加料线（q 线）方程可表示为：

$$y = \frac{q}{q-1} x - \frac{x_F}{q-1} \quad (3\text{-}6)$$

其中，

$$q = 1 + \frac{c_{pF}(t_S - t_F)}{r_F} \quad (3\text{-}7)$$

式中，q 为进料热状况参数；r_F 为进料液组成下的汽化潜热(kJ/kmol)；t_S 为进料液的泡点温度(℃)；t_F 为进料液温度(℃)；c_{pF} 为进料液在平均温度 $(t_S - t_F)/2$ 下的比热容(kJ/(kmol℃))；x_F 为进料液组成，摩尔分数。

回流比 R 的确定：

$$R = \frac{L}{D} \quad (3\text{-}8)$$

式中，L 为回流液量(kmol/s)；D 为馏出液量(kmol/s)。

式(3-8)只适用于泡点下回流时的情况，而实际操作时为了保证上升气流能完全冷凝，冷却水量一般都比较大，回流液温度往往低于泡点温度，即冷液回流。如图 3-1 所示，从全凝器出来的温度为 t_R、流量为 L 的液体回流进入塔顶第一块板，由于回流温度低于第一块塔板上的液相温度，离开第一块塔板的一部分上升蒸汽将被冷凝成液体，这样，塔内的实际流量将大于塔外回流量。

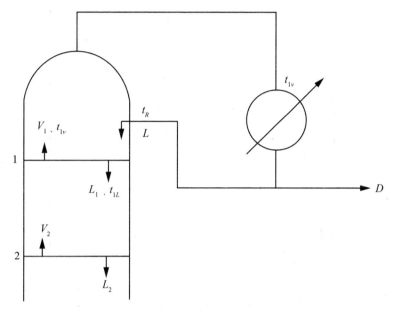

图 3-1 塔顶回流示意图

对第一块板作物料、热量衡算：

$$V_1 + L_1 = V_2 + L \tag{3-9}$$

$$V_1 I_{V1} + L_1 I_{L1} = V_2 I_{V2} + L I_L \tag{3-10}$$

对式(3-9)、式(3-10)整理、化简后，近似可得：

$$L_1 \approx L\left[1 + \frac{c_p(t_{1L} - t_R)}{r}\right] \tag{3-11}$$

即实际回流比为：

$$R_1 = \frac{L_1}{D} \tag{3-12}$$

$$R_1 = \frac{L\left[1 + \dfrac{c_p(t_{1L} - t_R)}{r}\right]}{D} \tag{3-13}$$

式中，V_1、V_2 为离开第1、第2块板的气相摩尔流量(kmol/s)；L_1 为塔内实际液流量(kmol/s)；I_{V1}、I_{V2}、I_{L1}、I_L 为别为对应 V_1、V_2、L_1、L 下的焓值(kJ/kmol)；r 为回流液组成下的汽化潜热(kJ/kmol)；c_p 为回流液在 t_{1L} 与 t_R 平均温度下的平均比热容(kJ/(kmol·℃))。

1) 全回流操作

在精馏全回流操作时，操作线在 x-y 图上为对角线，如图3-2所示，根据塔顶、塔釜的组成在操作线和平衡线间作梯级，即可得到理论塔板数。

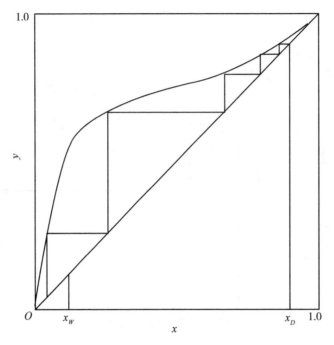

图3-2 全回流时理论塔板数的确定

2）部分回流操作

部分回流操作时，如图3-3所示，图解法的主要步骤为：

① 根据物系和操作压力在 x-y 图上作出相平衡曲线，并画出对角线作为辅助线；

② 在 x 轴上定出 $x=x_D$、x_F、x_W 三点，依次通过这三点作垂线分别交对角线于点 a、f、b；

③ 在 y 轴上定出 $y_C=\dfrac{x_D}{R+1}$ 的点 c，连接 a、c 作出精馏段操作线；

④ 由进料热状况求出 q 线的斜率 $\dfrac{q}{q-1}$，过点 f 作出 q 线交精馏段操作线于点 d；

⑤ 连接点 d、b 作出提馏段操作线；

⑥ 从点 a 开始在平衡线和精馏段操作线之间画阶梯，当梯级跨过点 d 时，就改在平衡线和提馏段操作线之间画阶梯，直至梯级跨过点 b 为止；

⑦ 所画的总阶梯数就是全塔所需的理论塔板数（包含再沸器），跨过点 d 的那块板就是加料板，其上的阶梯数为精馏段的理论塔板数。

3.1.3 实验装置

本实验装置的主体设备是筛板精馏塔，配套的有加料系统、回流系统、产品出料管

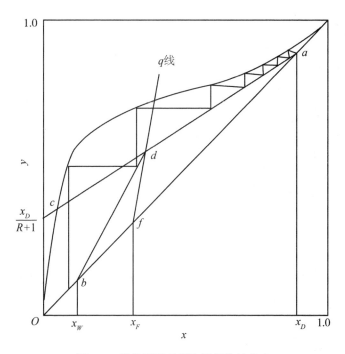

图 3-3　部分回流时理论塔板数的确定

路、残液出料管路、进料泵和一些测量、控制仪表。筛板塔主要结构参数：塔内径 $D=68\text{mm}$，厚度 $\delta=2\text{mm}$，塔节 $\phi 76\times 4$，塔板数 $N=10$ 块，板间距 $H_T=100\text{mm}$。加料位置由下向上起数第 4 块和第 6 块。降液管采用弓形，齿形堰，堰长 56mm，堰高 7.3mm，齿深 4.6mm，齿数 9 个。降液管底隙 4.5mm。筛孔直径 $d_0=1.5\text{mm}$，正三角形排列，孔间距 $t=5\text{mm}$，开孔数为 74 个。塔釜为内电加热式，加热功率为 2.5 kW，有效容积为 10 L。塔顶冷凝器、塔釜换热器均为盘管式。单板取样为自下而上第 1 块和第 10 块，斜向上为液相取样口，水平管为气相取样口。本实验料液为乙醇水溶液，釜内液体由电加热器产生蒸汽逐板上升，经与各板上的液体传质后，进入盘管式换热器壳程，冷凝成液体后再从集液器流出，一部分作为回流液从塔顶流入塔内，另一部分作为产品馏出，进入产品贮罐；残液经釜液转子流量计流入釜液贮罐。精馏过程如图 3-4 所示。

3.1.4　实验仿真装置

① 筛板精馏塔：精馏塔采用筛板结构，塔身用直径 $\phi 57\times 3.5\text{mm}$ 的不锈钢管制成，设有两个进料口，共 15 块塔板，塔板用厚度 1mm 的不锈钢板，板间距为 10cm；板上开孔率为 4%，孔径是 2mm，孔数为 21；孔按正三角形排列；降液管为 $\phi 14\times 2\text{mm}$ 的不锈钢管；塔高是 10mm；在塔顶和灵敏板的塔段中装有 WZG-001 微型铜电阻感温计各一支，并由仪表柜的 XCZ-102 温度指示仪加以显示。

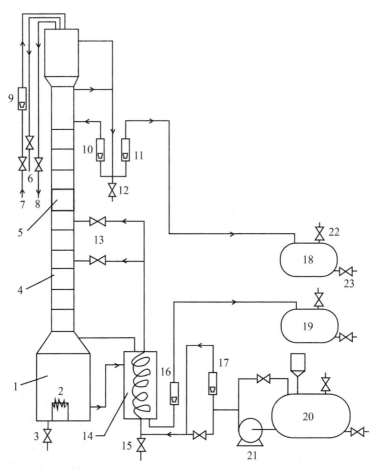

1：塔釜；2：电加热器；3：塔釜排液口；4：塔节；5：玻璃视镜；6：不凝性气体出口；7：冷却水进口；8：冷却水出口；9：冷却水流量计；10：塔顶回流流量计；11：塔顶出料液流量计；12：塔顶出料取样口；13：进料阀；14：换热器；15：进料液取样口；16：塔釜残液流量计；17：进料液流量计；18：产品灌；19：残液灌；20：原料灌；21：进料泵；22：排空阀；23：排液阀

图 3-4 筛板塔精馏塔实验装置图

② 蒸馏釜为 $\phi 250\times 340\times 3mm$ 不锈钢材质立式结构，用两支 1kW 的 SRY-2-1 型电热棒进行加热，其中一支为恒温加热，另一支则用自耦变压器调节控制，并由仪表柜上的电压、电流表加以显示。釜上有温度计和压力计，以测量釜内的温度和压力。

③ 冷凝器：采用不锈钢蛇管式冷凝器，蛇管规格为 $\phi 14\times 2mm$、长 2500mm，用自来水作冷却剂，冷凝器上方装有排气悬塞。

④ 产品贮槽：产品贮槽规格为 $\phi 250\times 340\times 3mm$，不锈钢材料制造，贮槽上方设有观察罩，以观察产品流动情况。

本实验进料的溶液为乙醇-水体系，其中乙醇占 20%（摩尔百分比）。溶液在储液罐中储备,用泵对塔进行进料,塔釜用电热器加热,电热器的电压由控制台来调整。塔釜的蒸汽到塔顶后,由塔顶的冷却器进行冷却(在仿真实验中设置为"常开",无需开关冷却水阀),冷却后的冷凝液进入储液罐,用回流的阀门及产品收集罐的阀门开度来控制回流比。产品进入产品收集罐。塔的压力由衡压调节阀来调节(在塔压高的时候可打开阀门进行降压,一般塔压控制在 1.2 atm 以下)。图 3-5 是筛板精馏仿真实验界面,图 3-6 是控制台的画面,图中标出了各种仪表及开关的名称。

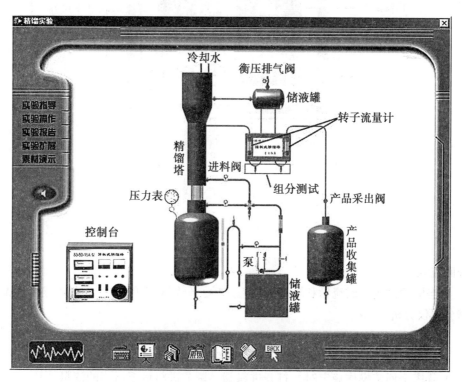

图 3-5　筛板精馏仿真实验界面

3.1.5　实验步骤

(1)全回流

① 配制浓度为 10%~20%(体积百分比)的料液加入贮罐中,打开进料管路上的阀门,由进料泵将料液打入塔釜,观察塔釜液位计高度,进料至釜容积的 2/3 处。进料时可以打开进料旁路的闸阀,加快进料速度。

② 关闭塔身进料管路上的阀门,启动电加热管电源,逐步增加加热电压,使塔釜温度缓慢上升(因塔中部玻璃部分较为脆弱,若加热过快玻璃极易碎裂,使整个精馏塔报废,故升温过程应尽可能缓慢)。

③ 打开塔顶冷凝器的冷却水,调节合适冷凝量,并关闭塔顶出料管路,使全塔处

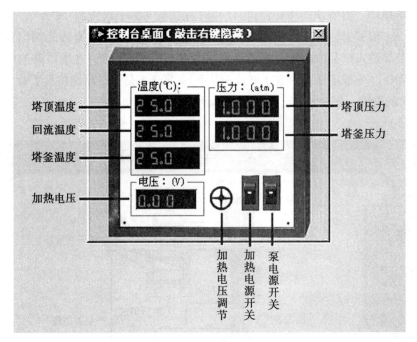

图 3-6 控制台的界面

于全回流状态。

④ 当塔顶温度、回流量和塔釜温度稳定后,分别取塔顶浓度 X_D 和塔釜浓度 X_W,送色谱分析仪分析。

(2) 部分回流

① 在储料罐中配制一定浓度的乙醇-水溶液(10%~20%)。

② 待塔全回流操作稳定时,打开进料阀,调节进料量至适当的流量。

③ 控制塔顶回流和出料两转子流量计,调节回流比 R($R=1~4$)。

④ 打开塔釜残液流量计,调节至适当流量。

⑤ 当塔顶、塔内温度读数以及流量都稳定后即可取样。

(3) 取样与分析

① 进料,塔顶、塔釜从各相应的取样阀放出。

② 塔板取样用注射器从所测定的塔板中缓缓抽出,取 1ml 左右注入事先洗净烘干的针剂瓶中,并给该瓶盖标号以免出错,各个样品尽可能同时取样。

③ 将样品进行色谱分析。

(5) 注意事项

① 塔顶放空阀一定要打开,否则容易因塔内压力过大导致危险。

② 料液一定要加到设定液位 2/3 处方可打开加热管电源,否则塔釜液位过低会使电加热丝露出干烧致坏。

③ 如果实验中塔板温度有明显偏差，是由于所测定的温度不是气相温度，而是气液混合的温度。

3.1.6 仿真实验步骤

① 第一步：全回流进料。

a. 打开泵开关。在控制台上用鼠标左键点击泵电源开关的上端（带白点的一端），打开泵电源开关，如图 3-7 所示。

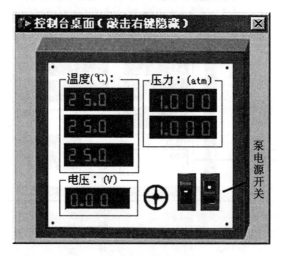

图 3-7 控制台泵电源开关位置

b. 打通进料的管线，依次打开阀门 1、2、3，向塔釜进料，进料至液位计的红点（正常液位标志）（屏幕显示为红点，此处显示为黑点）位置，完成进料，如图 3-8 所示。

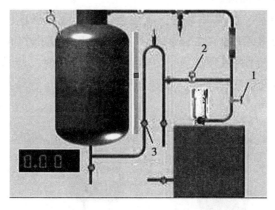

图 3-8 管线位置示意图

② 第二步：塔釜加热升温。

全回流进料完成后，开始加热，如图3-9所示。首先点击加热电源开关上端，打开加热电源开关。用鼠标点击加热电压调节手柄，左键增加电压，每点击一次加5V，右键减少电压，每点击一次减5V。或者在电压显示栏内用左键点击一下，输入所需的电压(0~350V)，然后在控制台窗口的空白处左键点击即可完成输入。

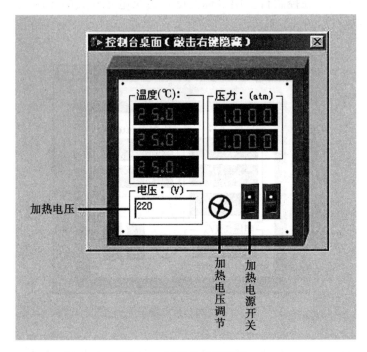

图3-9　加热位置示意图

③ 第三步：建立全回流。

a. 恒压加热开始后，回流开始前，应注意塔釜温度和塔顶压力的变化。当塔顶压力超过一个大气压很多时(如0.1 atm以上)，应打开恒压排气阀进行排气降压，如图3-10所示。此时应密切注视塔顶压力，当降到一个大气压时，应马上关闭。注意：回流开始以后就不能再打开恒压排气阀，否则会影响结果。

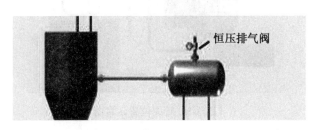

图3-10　恒压排气阀位置示意图

b. 塔顶的冷却水默认全开，当塔釜温度达到91℃左右时，开始有冷凝液出现（在塔顶及储液罐之间有细线闪烁）。此时鼠标左键点击回流支路上的转子流量计，如图3-11所示。鼠标左键点击转子流量计上的流量调节旋钮，左键增加，右键减少。也可以在开度显示框内填入所需的开度（0~100，百分比），然后在流量计上左键点击即可。调节阀的开度到100，开始全回流。

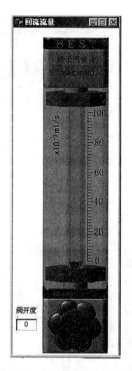

图3-11 转子流量计示意图

④ 第四步：读取全回流数据。

鼠标左键点击"组分测试"可看到组分含量（真实实验用仪器检测，此处简化），如图3-12所示。开始全回流10分钟以上，组分基本稳定达到正常值。当组分稳定以后，鼠标左键点击主窗口左侧菜单"数据处理"，在"原始数据"页填入数据（方法详见标准数据库操作方法），也可以使用自动记录功能进行记录。

⑤ 第五步：逐步进料，开始部分回流。逐渐打开塔中部的进料阀和塔底的排液阀以及产品采出阀，注意维持塔的物料平衡、塔釜液位和回流比，如图3-13所示。

⑥ 第六步：记录部分回流数据，参考记录全回流数据部分，将数据处理中的数据填好。注意事项：

a. 简化掉了配液过程，原料液直接装在原料罐内；

b. 加热电源开关由两个简化为一个；

c. 加热开始后，回流开始前，应注意塔釜温度和塔顶压力的变化。当塔顶压力超

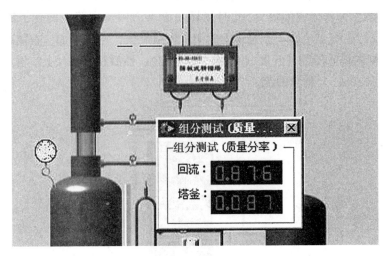

图 3-12　组分测试数据示意图

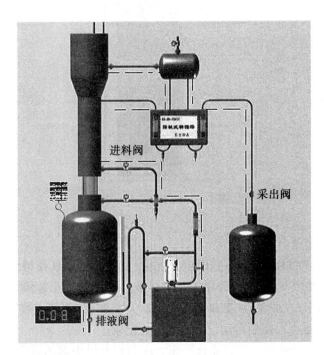

图 3-13　进料示意图

过一个大气压很多时(例如 0.1 atm 以上),应打开恒压排气阀进行排气降压。此时应密切注意塔顶压力,当降到一个大气压时,应马上关闭。注意:回流开始以后就不能再打开恒压排气阀,否则会影响结果。

d. 对于产品的检验，有些学校使用比重计，有些学校使用折光仪，各不相同，仿真实验中为了简化，我们直接给出了产品的摩尔分率。

3.1.7 数据记录及处理

① 将塔顶、塔底温度和组成，以及各流量计读数等原始数据列表；
② 按全回流和部分回流分别用图解法计算理论板数；
③ 计算全塔效率和单板效率；
④ 分析并讨论实验过程中观察到的现象。

3.1.8 仿真实验记录及处理

全回流和部分回流的数据处理基本相同。在原始数据处可看到自动记录的数据（或手工记录后填写的数据）。在计算结果处可看到自动计算的结果，也可以把手工计算的结果填入数据栏中（可由此数据画出特性曲线）。在理论板数项中可由计算结果中的数据画出精馏塔的特性曲线，如图3-14所示。

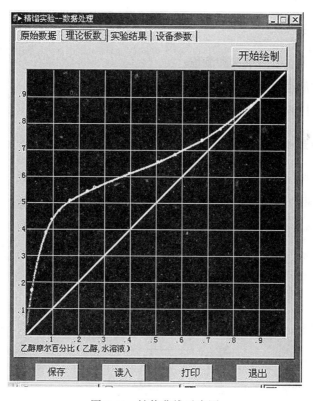

图3-14 性能曲线示意图

3.1.9 问题讨论

① 测定全回流和部分回流总板效率与单板效率时各需测几个参数？取样位置在何处？

② 全回流时测得板式塔上第 n、第 $n-1$ 层液相组成后，如何求得 x_n^*，部分回流时，又如何求得 x_n^*？

③ 在全回流时，测得板式塔上第 n、第 $n-1$ 层液相组成后，能否求出第 n 层塔板上的以气相组成变化表示的单板效率？

④ 进料液的汽化潜热时定性温度取何值？

⑤ 若测得单板效率超过 100%，作何解释？

⑥ 试分析实验结果成功或失败的原因，并提出改进意见。

3.2 填料吸收塔

3.2.1 实验目的

① 了解填料吸收塔的结构和流程；

② 了解吸收剂进口条件的变化对吸收操作结果的影响；

③ 掌握气相体积总传质系数的测定方法。

3.2.2 实验原理

填料塔中两相传质主要在填料有效湿表面上进行，需计算完成一定吸收任务所需的填料高度。由气液两相的平衡关系服从亨利定律，即平衡线为直线，操作线也为直线。

$$y_1 = \frac{p^*}{p} \tag{3-14}$$

$$\lg p^* = 7.024 - \frac{1161}{224 + T} \tag{3-15}$$

$$G(y_1 - y_2) = L(x_1 - x_2) \tag{3-16}$$

$$\Delta y_m = \frac{y_1 - mx_1 - (y_2 - 0)}{\ln \dfrac{y_1 - mx_1}{y_2 - 0}} \tag{3-17}$$

$$N_{OG} = \frac{y_1 - y_2}{\Delta y_m} \tag{3-18}$$

$$z = H_{OG} N_{OG} \tag{3-19}$$

$$H_{OG} = \frac{G}{K_{Y_a}\Omega} \tag{3-20}$$

式中,y_1 为气相进塔摩尔分数(塔底);y_2 为气相出塔摩尔分数(塔顶);x_1 为液相出塔摩尔分数(塔底);x_2 为液相进塔摩尔分数(塔顶);G 为气相摩尔流量(mol/s);L 为液相摩尔流量(mol/s);z 为填料高度(m);Δy_m 为气相对数平均浓度差;H_{OG} 为总传质单元高度(m);N_{OG} 为总传质单元数;m 为相平衡常数;K_{Y_a} 为气相体积总传质系数;Ω 为塔横截面积(m²);T 为液相温度(K)。

3.2.3 实验装置

由空气压缩机提供空气,经压力定值器定值为 0.02MPa,并经转子流量计计量后进入内盛丙酮的汽化器,产生丙酮和空气的混合气,混合气由塔底进入填料塔。在塔内同自塔顶喷下的水逆流接触,被吸收掉大部分丙酮后,从塔顶气体出口排出。由恒压高位槽底部流出的水,经转子流量计计量后,流经电加热器,由塔顶喷入吸收塔,吸收了空气中丙酮后由塔底经液封装置排入吸收液储罐,如图 3-15 所示。吸收塔参数见表 3-1,恒压槽参数见表 3-2,流量计参数见表 3-3。

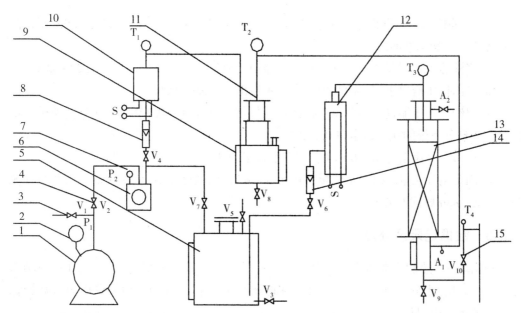

1:空气压缩机;2:压力表;3:空气压缩机旁路阀;4:空气压力调节阀;5:液体恒压槽;
6:气动压力定值器;7:压力表;8:空气流量计;9:丙酮汽化器;10:空气加热器;
11:丙酮蒸汽-空气混合器;12:水预热器;13:填料吸收塔;14:转子流量计;
15:液封;T_1、T_2、T_3、T_4:温度计;V_4、V_6、V_{10}:流量调节阀;V_3、V_5、V_7、V_8、V_9、V_{11}:启闭阀;A_1、A_2:气体进出口取样口

图 3-15 填料吸收实验装置图

表 3-1　　　　　　　　　　　吸收塔参数

塔径(mm)	塔身高度(mm)	填料名称	填料高度(mm)	填料尺寸(mm)
$\phi 41 \times 3$	500	瓷质拉西环	400	$6 \times 6 \times 1$

表 3-2　　　　　　　　　　　恒压槽参数

尺寸(mm)	吸液管在槽中插入深度(mm)
$\phi 300 \times 410$	370

表 3-3　　　　　　　　　　　流量计参数

空气转子流量计		液体转子流量计	
型号	流量范围（L/h）	型号	流量范围（L/h）
LZJ-6	100~1000	LZB-4	1~10

3.2.4　实验步骤

① 分别配置体积分数为 0.0%、1.0%、2.0%、4.0%、6.0% 的丙酮标准溶液，测定折光率，绘制标准曲线。

② 将液体丙酮用漏斗加入到丙酮汽化器中，液位高度为液位计高度的 2/3 以上。

③ 关闭恒压槽出水阀，向其中送水至不溢出，关闭进水阀。

④ 启动空气压缩机，调节压缩机使包内气体达到 0.05~0.1MPa，调节气动压力定值器，使进入系统的压力恒定在 0.02MPa。

⑤ 调节空气流量为 40~60L/h，调节水流量为 4~8 L/h。

⑥ 固定空气流量，改变水流量，恒定 5min 后读取数据；固定水流量，改变空气流量，恒定 5min 后读取数据。

3.2.5　数据记录及处理

根据表 3-4 绘制标准曲线并计算三个样品的摩尔分数。水密度_____g/L；丙酮密度_____g/L；水分子量_____g/mol；丙酮分子量_____g/mol。相关数据填入数据记录表和处理表中(见表 3-5、表 3-6)。

表 3-4　　　　　　　　　标准曲线及样品丙酮含量测定

样品名称	体积分数(%)	摩尔分数(%)	折光率
标液 1			

续表

样品名称	体积分数(%)	摩尔分数(%)	折光率
标液 2			
标液 3			
标液 4			
标液 5			
样品 1			
样品 2			
样品 3			

表 3-5　　　　　　　　　　数据记录表

序号	气相温度(K)	液相温度(K)	Q_G(L/h)	Q_L(L/h)	$p_{表}$(MPa)
1					
2					
3					

表 3-6　　　　　　　　　　数据处理表

序号	x_1 %	p^* (kPa)	y_1 %	G (mol/s)	L (mol/s)	y_2 %	m	Δy_m	N_{OG}	H_{OG} (m)	$K_Y a$ (mol·m^3·s)
1											
2											
3											

3.2.6　问题讨论

① 工业上，为什么吸收在低温、加压下进行，而解吸在高温、常压下进行？
② 讨论一下气相总体积传质系数 K_{Ya} 与 G 和 L 之间的关系。

3.3　流化床干燥实验

3.3.1　实验目的

① 了解流化床干燥装置的基本结构、工艺流程和操作方法；

② 学习测定物料在恒定干燥条件下干燥特性的实验方法；

③ 掌握根据实验干燥曲线求取干燥速率曲线以及恒速阶段干燥速率、临界含水量、平衡含水量的实验分析方法；

④ 实验研究干燥条件对于干燥过程特性的影响。

3.3.2 实验原理

在设计干燥器的尺寸或确定干燥器的生产能力时，被干燥物料在给定干燥条件下的干燥速率、临界湿含量和平衡湿含量等干燥特性数据是最基本的技术依据参数。由于实际生产中被干燥物料的性质千变万化，因此对于大多数具体的被干燥物料而言，其干燥特性数据常常需要通过实验测定而取得。按干燥过程中空气状态参数是否变化，可将干燥过程分为恒定干燥条件操作和非恒定干燥条件操作两大类。若用大量空气干燥少量物料，则可以认为湿空气在干燥过程中温度、湿度均不变，再加上气流速度以及气流与物料的接触方式不变，则称这种操作为恒定干燥条件下的干燥操作。

(1) 干燥速率的定义

干燥速率定义为单位干燥面积(提供湿分汽化的面积)、单位时间内所除去的湿分质量，即

$$U = \frac{dW}{Ad\tau} = -\frac{G_C dX}{Ad\tau} \quad kg/(m^2 \cdot s) \tag{3-21}$$

式中，U 为干燥速率，又称干燥通量($kg/(m^2 \cdot s)$)；A 为干燥表面积(m^2)；W 为汽化的湿分量(kg)；τ 为干燥时间(s)；G_C 为绝干物料的质量(kg)；X 为物料湿含量(kg 湿分/kg 干物料)，负号表示 X 随干燥时间的增加而减少。

(2) 干燥速率的测定方法

① 方法一：

a. 将电子天平开启，待用；

b. 将快速水分测定仪开启，待用；

c. 准备 0.5~1kg 的湿物料，待用；

d. 开启风机，调节风量至 40~60m³/h，打开加热器加热。待热风温度恒定后(通常可设定在 70~80℃)，将湿物料加入流化床中，开始计时，每过 4min 取出 10 克左右的物料，同时读取床层温度。将取出的湿物料在快速水分测定仪中测定，得初始质量 G_i 和终了质量 G_{iC}。则物料中瞬间含水率 X_i 为：

$$X_i = \frac{G_i - G_{iC}}{G_{iC}} \tag{3-22}$$

② 方法二(数字化实验设备可用此法)：利用床层的压降来测定干燥过程的失水量。

a. 准备 0.5~1kg 的湿物料，待用；

b. 开启风机，调节风量至 40~60m³/h，打开加热器加热。待热风温度恒定后(通常可设定在 70~80℃)，将湿物料加入流化床中，开始计时，此时床层的压差将随时间减

小，实验至床层压差（Δp_e）恒定为止。则物料中瞬间含水率X_i为：

$$X_i = \frac{\Delta p - \Delta p_e}{\Delta p_e} \tag{3-23}$$

式中，Δp为τ时刻床层的压差。

计算出每一时刻的瞬间含水率X_i，然后将X_i对干燥时间τ_i作图，如图3-16所示，即为干燥曲线。

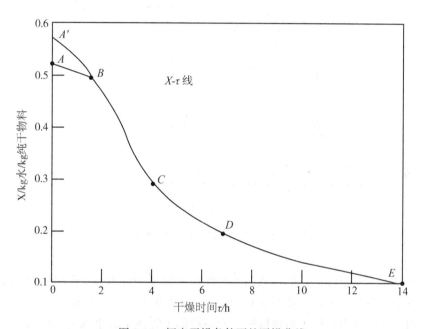

图3-16 恒定干燥条件下的干燥曲线

上述干燥曲线还可以变换得到干燥速率曲线。由已测得的干燥曲线求出不同X_i下的斜率$\dfrac{\mathrm{d}X_i}{\mathrm{d}\tau_i}$，再由式（3-21）计算得到干燥速率$U$，将$U$对$X$作图，就是干燥速率曲线，如图3-17所示。

将床层的温度对时间作图，可得床层的温度与干燥时间的关系曲线。

（3）干燥过程分析

1）预热段

见图3-16、图3-17中的AB段或$A'B$段。物料在预热段中，含水率略有下降，温度则升至湿球温度t_w，干燥速率可能呈上升趋势变化，也可能呈下降趋势变化。预热段经历的时间很短，通常在干燥计算中忽略不计，有些干燥过程甚至没有预热段。

2）恒速干燥阶段

恒速干燥阶段见图3-16、图3-17中的BC段。该段物料水分不断汽化，含水率不断下降。但由于这一阶段去除的是物料表面附着的非结合水分，水分去除的机理与纯水的

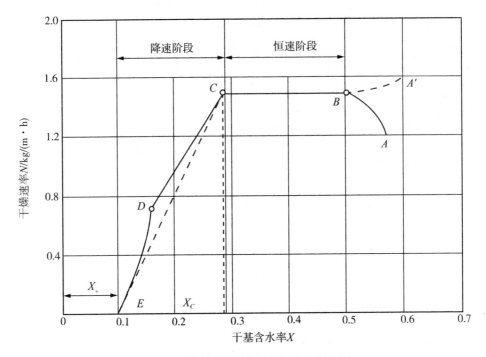

图 3-17 恒定干燥条件下的干燥速率曲线

相同,故在恒定干燥条件下,物料表面始终保持为湿球温度 t_w,传质推动力保持不变,因而干燥速率也不变。于是,在图 3-17 中,BC 段为水平线。只要物料表面保持足够湿润,物料的干燥过程中总处于恒速阶段。而该段的干燥速率大小取决于物料表面水分的汽化速率,亦即决定于物料外部的空气干燥条件,故该阶段又称为表面汽化控制阶段。

3) 降速干燥阶段

随着干燥过程的进行,物料内部水分移动到表面的速度赶不上表面水分的气化速率,物料表面局部出现"干区",尽管这时物料其余表面的平衡蒸汽压仍与纯水的饱和蒸汽压相同,但以物料全部外表面计算的干燥速率因"干区"的出现而降低,此时物料中的含水率称为临界含水率,用 X_c 表示,对应图 3-17 中的 C 点,称为临界点。过 C 点以后,干燥速率逐渐降低至 D 点,C 至 D 阶段称为降速第一阶段。干燥到点 D 时,物料全部表面都成为干区,汽化面逐渐向物料内部移动,汽化所需的热量必须通过已被干燥的固体层才能传递到汽化面;从物料中汽化的水分也必须通过这一干燥层才能传递到空气主流中。干燥速率因热、质传递的途径加长而下降。此外,在点 D 以后,物料中的非结合水分已被除尽。接下去所汽化的是各种形式的结合水,因而,平衡蒸汽压将逐渐下降,传质推动力减小,干燥速率也随之较快降低,直至到达点 E 时,速率降为零。这一阶段称为降速第二阶段。降速阶段干燥速率曲线的形状随物料内部的结构而异,不一定都呈现出前面所述的曲线 CDE 形状。对于某些多孔性物料,可能降速两个阶段的界限不是很明显，曲线好像只有 CD 段；对于某些无孔性吸水物料，汽化只在表面进行，干燥速率取决于固体内部水分的扩散速率，故降速阶段只有类似 DE 段的曲线。与恒速阶段相比，降速阶段从物料中除去的水分量相对少许多，但所需的干燥时间却长得多。总之，降速阶段的干燥速率取决于物料本身的结构、形状和尺寸，而与干燥介质状况关系不大，故降速阶段又称物料内部迁移控制阶段。

3.3.3 实验装置

（1）装置流程

本装置流程如图 3-18 所示。

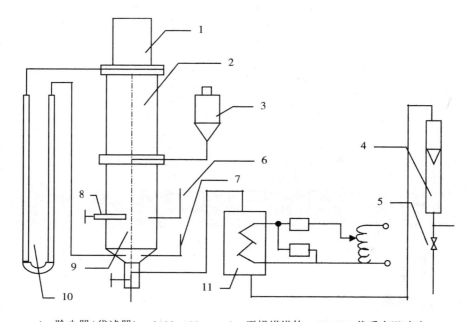

1：除尘器（袋滤器），φ130×120mm；2：干燥塔塔体，φ146×8 优质高温玻璃；3：加水器，0~400mL；4：气体转子流量计，LZB-25 0~25m³/n；5：流量调节阀；6：温度计，0~150℃ CU50 铜电阻；7：温度计，0~150℃ CU50 铜电阻；8：固体物料取样器，2.3 克/次；9：实验用干燥物料，30~40 目变色硅胶；10：压差计，±50cm 水银；11：电加热器，3kW

图 3-18 实验装置与流程图

（2）主要设备及仪器

① 鼓风机：BYF7122，370W；
② 电加热器：额定功率 2.0kW；
③ 干燥室：φ100mm×750mm；
④ 干燥物料：耐水硅胶；
⑤ 床层压差：Sp0014 型压差传感器或 U 形压差计。

3.3.4 实验仿真装置

实验仿真装置界面如图 3-19 所示。

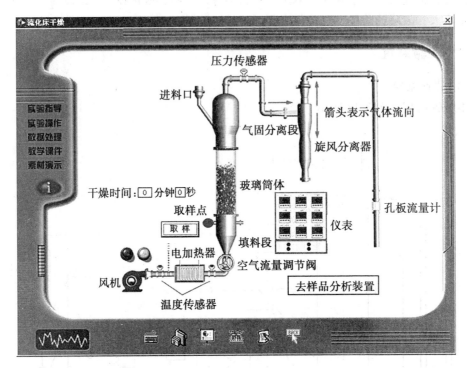

图 3-19 实验仿真装置界面

3.3.5 实验步骤

① 接通气源并缓慢调节风量使干燥塔中颗粒物料处于良好的流化状态(注意压差计读数,勿使测压指示液冲出);
② 向加水器中加入适量的水,调节加水器下部铜旋塞,勿使注入干燥塔的水流速度过大,加水时应使取样器保持拉出位置,同时塔内处于流化状态;
③ 开通风源,打开阀门5,调节空气流量,接通电源,在智能温度调节仪 AI-708 上设定控制温度 95~100℃;
④ 在气体的流量和温度维持一定的条件下,每隔一定时间记录床层温度,并取样分析固体物料的含水量;
⑤ 固体物料取样时只要把取样器推入,随即拉出即可;
⑥ 实验进行直至物料温度明显升高,硅胶变蓝即可停止;
⑦ 实验停止步骤:切断电源,待气体温度下降后,停止送风;
⑧ 当塔中需要补充硅胶物料时,卸下袋滤器后可加入;

⑨ 当更换硅胶物料时，可用吸尘器的皮管伸入塔体内即可全部吸出。

3.3.6 仿真实验步骤

（1）干燥实验过程

① 打开风机，开始实验。

② 先把空气流量调节阀打开到不小于 42 的开度，使系统能进入到流化床阶段。

③ 打开仪表面板的加热器开关(手动或者自动开关均可)。点击"自动记录"按钮，记录实验数据；也可手动记录数据，手动记录数据时，需同时点击"取样"按钮进行取样。

④ 以后每间隔 10 分钟左右记录一组数据，取至少 10 组以上数据，实验进行到后期，取样间隔时间可减少到六七分钟一次。主窗体上有时间显示。取样和记录实验数据在同一分钟内进行即可。

⑤ 本实验设计的干燥时间为 90 至 100 分钟，因此，实验进行到 100 分钟后即可停止，进入到样品分析装置。

⑥ 样品选取栏里存有每次取得的样品，可逐个选取样品，进行称量，每次称量完毕后，在电子称的显示屏幕上可读取样品质量，每次取样为 10g，均烘干后称量，称量所得质量为样品的净重，10g 减去样品的净重为样品所含水分的重量，用样品含的水分的质量除以 10g，得到样品的含水率。可手动记录所得样品的含水率，也可点击本窗口的"自动记录"按钮进行自动记录，自动记录先自动计算出样品的含水率，然后再记录到相应的数据表格里。

⑦ 样品分析完后点击主界面的"数据处理"按钮，进入到数据处理界面，选择干燥实验数据选项卡。拉横向滚动条到最后，一直到干燥速率栏，点击"计算干燥速率"按钮，计算出干燥速率。

⑧ 原始数据采集完毕，绘制坐标图。

⑨ 进入到"临界含水量和传质系数"选项卡，从干燥速率曲线上读取临界含水率的值，点击"计算传质系数"按钮，计算出传质系数。

（2）流化实验过程

① 关闭仪表窗体上的加热器开关(手动和自动开关全部关闭)。

② 将空气流量调节阀关闭，然后慢慢开大，适当改变阀门开度，并在流化实验数据选项卡中记录下相应阀门开度下的气速和床层压降，可手动记录，也可自动记录。由于空气流速增加到 1kPa 的时候进入到流化床阶段，此后床层压降即为定值，所以，在流速增加到 1kPa 之前，至少取 5~6 组数据，以便更好地拟合曲线。整个实验取不少于 10 组数据。

③ 实验原始数据记录完毕后，绘制曲线。点击"流化曲线"选项卡，然后点击"自动绘制"按钮，即画出流化曲线。

3.3.7 数据记录及处理

请将实验数据填入表3-7中,并完成以下任务:
① 绘制干燥曲线(失水量-时间关系曲线);
② 根据干燥曲线作干燥速率曲线;
③ 读取物料的临界湿含量;
④ 绘制床层温度随时间变化的关系曲线;
⑤ 对实验结果进行分析讨论。

实验条件:室内温度_____;室内相对湿度_____;气流量_____;热气温度计温度_____。

表 3-7　　　　　　　　　　数据记录表

序号	容器重量	湿料与容器重量	湿料重量	干料与容器重量	除去的水分重量	床层温度 ℃/10分钟	中间温度	压差计压力
1								
2								
3								
4								
5								
6								
7								
8								
9								
10								
11								
12								
13								

备注:每隔10分钟测试一个数据,取样物料放入烘箱中在120℃温度下烘烤1小时;天平1/1000精度。

3.3.8 问题讨论

① 什么是恒定干燥条件?本实验装置中采用了哪些措施来保持干燥过程在恒定干燥条件下进行?

② 控制恒速干燥阶段速率的因素是什么？控制降速干燥阶段干燥速率的因素又是什么？

③ 为什么要先启动风机，再启动加热器？实验过程中床层温度是如何变化的？为什么？如何判断实验已经结束？

④ 若加大热空气流量，干燥速率曲线有何变化？恒速干燥速率、临界湿含量又如何变化？为什么？

Chapter 1 Typical Momentum Transfer Experiment

1.1 Determination of Characteristic Curve of Centrifugal Pump

1.1.1 Purpose of the Experiment

① To understand the structure and characteristics of the centrifugal pump and be familiar with the method of its operation;

② To master the methods of determination of the characteristic curve of the centrifugal pump;

③ To understand the working principle and utilization of the electric regulating valve.

1.1.2 Principle of the Experiment

The characteristic curve of the centrifugal pump is an important basis for the selection and use of the centrifugal pump. It is the relation curve between the lift H, axial power N and efficiency η with the flow rate Q of the pump, which is the macroscopic manifestation of the movement law of the fluid in the pump. Due to the complexity of the flow situation inside the pump, the characteristic relation curve of the pump cannot be deduced theoretically, instead, it can only be determined by experimentation.

(1) Determination and Calculation of Lift H

Let the inlet vacuum gauge and the outlet pressure gauge of the centrifugal pump be section, thus the following mechanical energy balance equation:

$$z_1 + \frac{p_1}{\rho g} + \frac{u_1^2}{2g} + H = z_2 + \frac{p_2}{\rho g} + \frac{u_2^2}{2g} + \Sigma h_f \tag{1-1}$$

In view of the insignificant length of the pipe between the two sections, the resistance can usually be neglected and since the square difference of speed is also negligible, thus the following equation is valid:

$$H = (z_2 - z_1) + \frac{p_2 - p_1}{\rho g} = H_0 + H_1(表值) + H_2 \tag{1-2}$$

Wherein: $H_0 = z_2 - z_1$, meaning the potential difference between the outlet and inlet of the pump, unit: m.

Wherein, ρ: fluid density (kg/m^3); g: acceleration of gravity (m/s^2); p_1, p_2: respectively the vacuum degree and gauge pressure of the inlet and outlet of the pump (Pa); H_1, H_2: the pressure head responding to the vacuum degree and gauge pressure of the inlet and outlet of the pump, m; u_1, u_2: respectively the flow rate at the inlet and outlet of the pump (m/s); z_1, z_2: respectively the installation height of the vacuum gauge and pressure gauge (m).

From the above equation, it can be known that, with the reading of the vacuum gauge and pressure gauge, as well as the difference between the installation heights of the two gauges, the lift of the pump can be calculated.

(2) Measurement and Calculation of Axial Power N

$$N = N_{electron} \cdot k \ (W) \tag{1-3}$$

Wherein, the power is the value displayed on the electric power gauge, and k represents motor drive efficiency, whose value is set at 0.95.

(3) Calculation Efficiency η

The efficiency η of the pump is the ratio between the effective power Ne and the axial power N of the pump. The effective power Ne is the actual power gained when the fluid passes through the pump in unit time. The axial power N is the power gained from the pump motor in unit time. The difference between the two reflects the magnitude of the hydraulic loss, volume loss and mechanical loss of the pump. The effective power Ne of the pump can be calculated using equation (1-4):

$$Ne = HQ\rho g \tag{1-4}$$

Therefore, the efficiency of the pump is:

$$\eta = \frac{HQ\rho g}{N} \times 100\% \tag{1-5}$$

(4) Conversion When Rotation Rate Changes

The characteristic curve of the pump is obtained experimentally at fixed rotation rate. However, in reality, when the torque of the induction motor changes, its rotation rate will also change, as a result of which, as the flow Q changes, the rotation rate n at multiple experimental points will be varied somewhat, so that before the characteristic curve can be plotted, the measured data must be converted to that at a certain speed n' (which can be 2900 rpm, the rated rotation rate of the centrifugal pump). The conversion goes as follows:

Flow Volume
$$Q' = Q \frac{n'}{n} \tag{1-6}$$

Lift
$$H' = H \left(\frac{n'}{n}\right)^2 \tag{1-7}$$

Chapter 1 Typical Momentum Transfer Experiment

Axial Power
$$N' = N\left(\frac{n'}{n}\right)^3 \tag{1-8}$$

Efficiency
$$\eta' = \frac{Q'H'\rho g}{N'} = \frac{QH\rho g}{N} = \eta \tag{1-9}$$

1.1.3 Experimental Devices

The measuring device of the characteristic curve of the centrifugal pump is indicated by Figure 1-1:

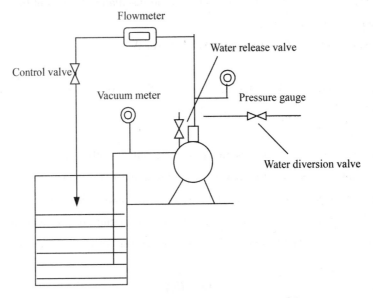

Figure 1-1 Schematic Diagram of the Experimental Device

1.1.4 Simulation Devices of the Experiment

The interface of the performance curve simulation experiment device of the centrifugal pump is shown in Figure 1-2: Wherein, the rotation rate of the pump: 2900 r/min; rated lift: 20 m; electric motor efficiency: 93%; transmission efficiency: 100%; water temperature: 25℃; pump inlet pipe inner diameter: 41 mm; pump outlet pipe inner diameter: 35.78 mm; vertical distance between the two pressure gauging port: 0.35m; flow coefficient of turbine flowmeter: 75.78.

1.1.5 Experimental Procedure

① Clean the water tank and add in the water for the experimental purpose. Prime the centrifugal pump using a priming funnel and expel the gas from the pump.

1.1 Determination of Characteristic Curve of Centrifugal Pump

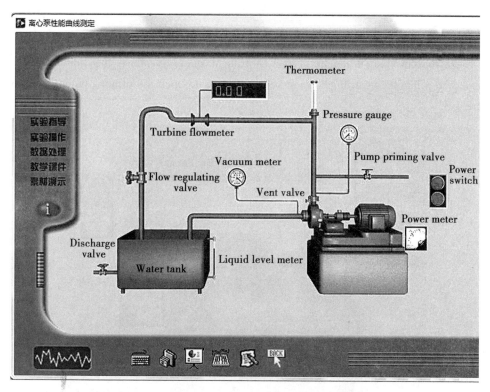

Figure 1-2 The Interface of the Performance Curve Simulation Experiment Device of the Centrifugal Pump

② Check the opening of each valve and the self-inspection status of the meters, and if the electric motor and the centrifugal pump operate normally during the trial run. Before turning on the centrifugal pump, close the outlet valve first. Only when the pump reaches its rated rotation rate can the outlet valve be opened one by one.

③ In the experiment, the outlet flow regulation gate valve is gradually opened to increase the flow rate, and the corresponding data can be read after the readings of each instrument are shown to be stable. The main data to be gained through the centrifugal pump characteristics experiment are: flow rate Q, pump inlet pressure p_1, pump outlet pressure p_2, electric motor power N, pump rotation rate n, fluid temperature t and the height difference between two pressure gauging points H_0 ($H_0 = 0.1 \text{m}$).

④ Change the opening of the outlet regulating valve, after obtaining about 10~15 data sets, stop the pump, and record the relevant data of the device (such as centrifugal pump model, its rated flow, rated rotation rate, lift and power, etc.). Stop the pump after closing the outlet flow regulating valve.

1.1.6 Procedure of Simulation Experiment

(1) Pump Priming

Since the installation height of the centrifugal pump is above the liquid level, it is necessary to prime the centrifugal pump before starting it. Open the priming pump, as shown in Figure 1-3.

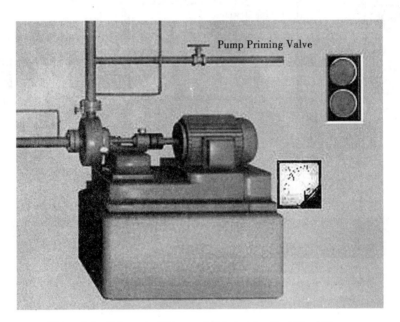

Figure 1-3   Schematic Diagram of Priming the Centrifugal Pump

Click the left button of the mouse on the pressure gauge to magnify the readings (click the right button to return). When the reading is greater than 0, it means the pump casing has been filled with water, but because the upper part of the pump casing still retains a small quantity of air, it is necessary to expel the air first. Turn up the opening of the exhaust valve to over 0 to expel the air. When the air is expelled off, some fluid will well, as shown in the Figure1-3. At this point, close the exhaust valve and the priming valve to terminate the priming.

(2) Turning on the Pump

After priming the pump, turn on the pump's power supply to start the centrifugal pump. Note: when starting the centrifugal pump, the main regulating valve should be closed. If the main regulating valve is fully opened, it will lead to excessive power when the pump starts, which may burn out the pump.

(3) Establishing the Flow

After starting the centrifugal pump, adjust the main regulating valve to be opened at 100, as shown in Figure 1-4.

1.1 Determination of Characteristic Curve of Centrifugal Pump

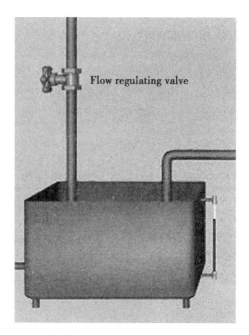

Figure 1-4　Flow Regulating Valve of the Centrifugal Pump

(4) Reading the Data

Read the data when the numbers displayed by the turbine flowmeter is stable. Click the left button of the mouse on the pressure gauge, vacuum gauge and power gauge to magnify the display and read the data, as shown in Figure 1-5.

Figure 1-5　Schematic Diagram of Gauges the Centrifugal Pump

Chapter 1 Typical Momentum Transfer Experiment

1.1.7 Data Recording and Processing

① Record the original data of the experiment, as shown in Table 1-1:

Centrifugal pump model: _____ ; rated flow rate: _____ ; rated lift: _____ ; rated power: _____ ; height difference between pressure gauging points at the inlet and outlet of the pump H_0: _____ ; fluid temperature t: _____

Table 1-1   **Experiment Record of the Centrifuge Pump**

No.	Flow Volume $Q(m^3/h)$	Inlet Pressure p_1 of Pump (kPa)	Outlet Pressure p_2 of Pump (kPa)	Power of Electric Motor $N_{electron}$ (kW)	Rotation Rate of Pump n (r/min)

② According to the equation listed in the Principle of Experiment section, after checking the rotation rate by the law of proportionality, calculate the lift, axial power and efficiency of the pump under each flow volume, as shown in Table 1-2. At the same time, it is necessary to plot respectively the H-Q, N-Q and η-Q curves at certain rotation rate; analyze the result of the experiment to determine the most appropriate working range of pump.

Table 1-2   **Data Processing Results**

No.	Flow Volume Q' (m^3/h)	Lift H' (m)	Axial Power N' (kW)	Pump Efficiency η' (%)

continued

No.	Flow Volume Q' (m³/h)	Lift H' (m)	Axial Power N' (kW)	Pump Efficiency η' (%)

1.1.8 Recording and Processing of Simulation Experiment

(1) Experimental Record

Click the left button of the mouse on the Data Processing item of the menu on the left side of the main experiment screen to call out the data processing window; in the original data page, fill in the data in the table by item. Or record the data in the data recording table printed by clicking the "print data recording table" key. The two forms are basically the same, but pay attention to the conversion of unit. The simulation experiment recording table is shown in Figure

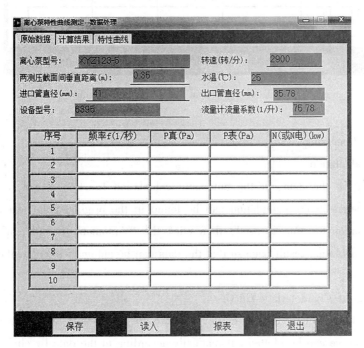

Figure 1-6 Blank Simulation Experiment Table of the Centrifugal Pump

1-6. Note: If the automatic recording function is used, when the "automatic record" key is clicked, the data will be automatically written with no need for manual typing. If the automatic recording function is used or the data is recorded in the database, then skip this step. If the data is recorded in the data recording table printed by clicking the Print Data Recording Table key, please fill the data in the table as shown in Figure 1-7.

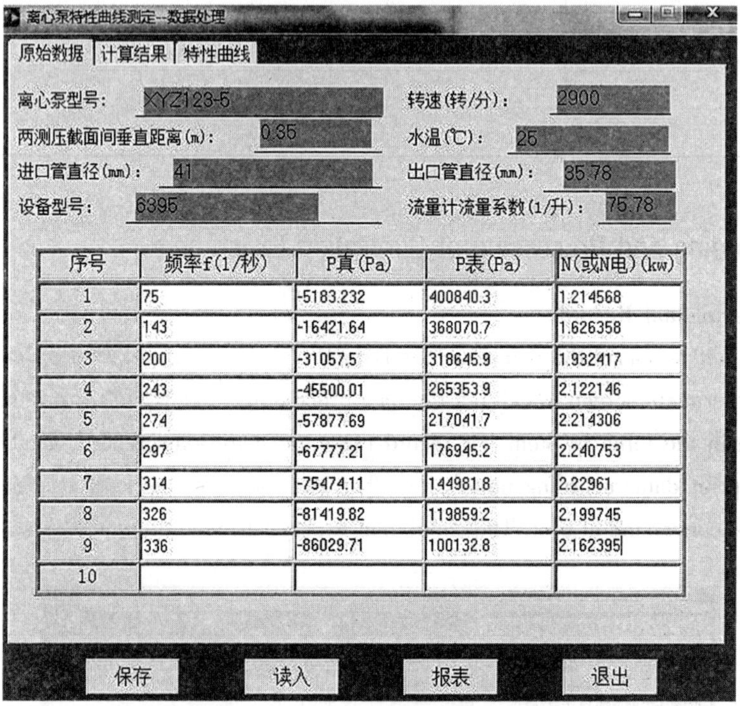

Figure 1-7 Simulation Experiment Data Table of the Centrifugal Pump

(2) Data Processing

1) Data Calculation

After filling in the data, if the automatic calculation function is not used, then the necessary parameters in the calculation can be found in the original data page, while if the automatic calculation function is used, then click the "Automatic Calculation" key on the corresponding calculation result page so that the data can be automatically calculated and filled in. See Figure 1-8.

2) Plotting the Characteristic Curve

The result of the characteristic curve is shown in Figure 1-9. After the calculation is completed, the curve can be plotted automatically according to the data by clicking the "Start Plotting" button in the curve page as shown in Figure 1-9. Click the "Print" button among a

1.1 Determination of Characteristic Curve of Centrifugal Pump

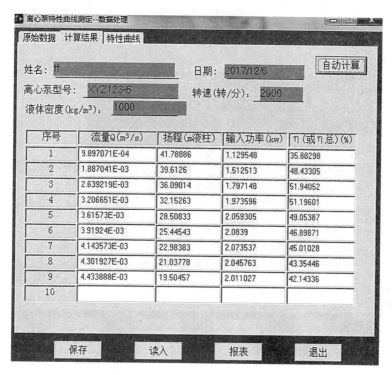

Figure 1-8 Data Calculation Result

row of buttons at the bottom of the data processing window to call out the experiment report window (see Figure 1-10). Click the "Save" button among a row of buttons at the bottom of the data processing window to save the raw data in a disk file, and click the "Load" button to read the file.

1.1.9 Discussion

① With the measured experiment data, try to analyze why the outlet valve has to be closed when the centrifugal pump is turned on.

② Why should the pump be primed before the centrifugal pump is started? If after the priming, the pump still fails to start, what do you think may be the cause?

③ Why should we use the pump's outlet valve to adjust the flow? What are the advantages and disadvantages of this method? Is there any other method to adjust the flow?

④ When the pump is started, if the outlet valve is not opened, will the readings of the pressure gauge rise gradually? Why?

⑤ Is it reasonable to install a valve on the inlet pipe of a normally working centrifugal pump? Why?

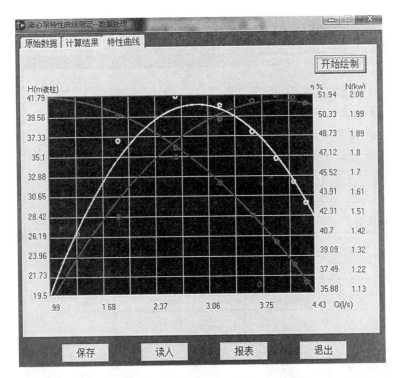

Figure 1-9 Characteristic Curve of Centrifugal Pump

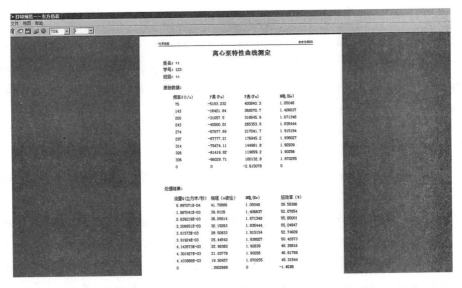

Figure 1-10 Experiment Report Window

⑥ Try to analyze, when a clean water pump is used to transport salt water of 1200 kg/m^3 concentration, do you think the pump pressure would change given the same flow rate? Would the axial power change?

1.2 Determination of Fluid Resistance in Pipe

1.2.1 Purpose of the Experiment

① To master the general experimental method for measuring the resistance loss when a fluid flows through a straight pipe, a pipe fitting or a valve.

② To determine the relationship between the friction coefficient λ of the straight pipe and the reynolds number Re, and verify the relation curve between λ and Re in the general turbulent region.

③ To determine the local resistance coefficient ξ when a fluid flows through a pipe or valve.

④ To grasp the method of using the inverted U type differential pressure gauge and the turbine flowmeter.

⑤ To identify the various pipe fittings and valves that make up the pipeline and understand their functions.

1.2.2 Principle of the Experiment

When a fluid passes through a pipeline system made up of straight pipes, pipe fittings (such as three-way connectors and elbows, etc.) and valves, due to the existence of viscous shear stress and eddy current stress, the mechanical energy of the fluid will inevitably be reduced. The mechanical energy loss caused when a fluid flows through a straight pipe is called the straight pipe resistance loss. The mechanical energy loss caused by the direction and/or velocity change of the fluid as it flows through a pipe or valve is called the local resistance loss.

(1) Determination of Straight Pipe Resistance and Friction Coefficient λ

When the fluid flows steadily in a horizontal equal-diameter straight pipe, the resistance loss is as follows:

$$w_f = \frac{\Delta p_f}{\rho} = \frac{p_1 - p_2}{\rho} = \lambda \frac{l}{d} \frac{u^2}{2} \tag{1-10}$$

That is,
$$\lambda = \frac{2d\Delta p_f}{\rho l u^2} \tag{1-11}$$

Wherein, λ: straight pipe resistance and friction coefficient, dimensionless; d: inner diameter of pipe(m); Δp_f: the pressure drop as the fluid flows through a straight pipe l meters long(Pa); w_f: the mechanical energy loss as a unit mass of fluid flows through a straight pipe l meters long(J/kg); ρ: fluid density (kg/m³); l: length of the straight pipe(m); u: the average velocity of fluid flowing in the pipe(m/s).

When in stagnation (static flow):

$$\lambda = \frac{64}{Re} \tag{1-12}$$

$$Re = \frac{du\rho}{\mu} \tag{1-13}$$

Wherein, Re: Reynolds number, dimensionless; μ: fluid viscosity(kg/(m·s)).

When in turbulence, λ is the function of the Reynolds number Re and the relative roughness (ε/d), which must be determined in an experiment. Judging from equation (1-11), it can be seen that the determination of λ, l and d need to be determined and parameters such as Δp_f, u, ρ and μ have to be measured. l and d are device parameters (given in the device parameter table); ρ and μ are obtained by measuring the fluid temperature and then referring to the relevant manual; u is obtained by measuring the flow volume of the fluid and then calculating the pipe diameter. The flow volume V is measured using a turbine flowmeter, m³/h.

$$u = \frac{V}{900\pi d^2} \tag{1-14}$$

Δp_f can be measured using a liquid column piezometer such as U pipe, inverted U pipe or pressure pipe, etc., or displayed using the differential pressure transmitter and secondary instrument. When an inverted U pipe liquid column piezometer is used,

$$\Delta p_f = \rho g R \tag{1-15}$$

Wherein, R: height of the water column(m).

When a U pipe liquid column piezometer is used,

$$\Delta p_f = (\rho_0 - \rho) g R \tag{1-16}$$

Wherein, R: height of the liquid column(m); ρ_0: density of the indicating liquid(kg/m³).

According to the experiment device structure parameters l and d, the density of the indicating liquid ρ_0, the temperature of the fluid t_0 (refer to physical properties ρ and μ of the fluid), and the flow volume V measured in the experiment and the reading R of the liquid column piezometer, calculate Re and λ through the equations of (1-14), (1-15) or (1-16), (1-11) and (1-13) before projecting the two on the log-log plot.

(2) Determination of Local Resistance Coefficient ξ

The local resistance loss can usually be expressed in two ways, namely the equivalent length method and the resistance coefficient method.

1) Equivalent Length Method

The mechanical energy loss caused when the fluid flows through a pipe or valve can be seen to be equivalent to the mechanical energy loss when the fluid flows through an equal-diameter pipe of certain length, which length is called the equivalent length and indicated with the symbol le. In this way, the local resistance loss can be calculated using formula of the straight pipe resistance, and moreover, in the calculation of a pipeline system, the length of the straight pipe in the system can be combined with the equivalent length of the pipes and valves thereof, so that the total mechanical energy loss $\sum w_f$ of the fluid flowing in the pipeline is:

$$\sum w_f = \lambda \frac{l + \sum l_e}{d} \frac{u^2}{2} \qquad (1\text{-}17)$$

2) Resistance Coefficient Method

The method which expresses the mechanical energy loss caused when a fluid flows through certain pipe or valve as a certain times of local resistance to the average kinetic energy of the fluid flowing in a small-section pipe is called the resistance coefficient method. That is:

$$w'_f = \frac{\Delta p'_f}{\rho} = \xi \frac{u^2}{2} \qquad (1\text{-}18)$$

Thus
$$\xi = \frac{2\Delta p'_f}{\rho u^2} \qquad (1\text{-}19)$$

Wherein, ξ: local resistance coefficient, dimensionless; $\Delta p'_f$: local pressure drop (Pa); (the measured pressure drop should deduct the pressure drop of straight pipe section between two pressure measuring ports, and the pressure drop of straight pipe section is obtained through the straight pipe resistance experiment). ρ: fluid density (kg/m³); g: acceleration of gravity, 9.81 m/s²; u: the average velocity of fluid in a small-section pipe (m/s).

The fittings and valves to be measured are designated on site. In this experiment, the resistance coefficient method is used to represent the local resistance loss of the pipe fitting or valve. According to the diameter d of the pipe connecting the two ends of the pipe or valve, the density ρ_0 of the indicating fluid, the fluid temperature t_0 (refer to the physical properties ρ and μ of the fluid), the flow volume V measured in the experiment, and the reading R of the liquid column piezometer, obtain the local resistance coefficient ξ of the pipe fitting or valve through the equations of (1-14), (1-15) or (1-16), (1-19).

1.2.3 Experimental Devices

(1) Experimental Device (See Figure 1-11)

(2) Experimental Process

The experiment setup is made up of the water storage tank, centrifugal pump, water pipes of different diameters and materials, valves of various sorts, fittings, turbine flowmeter and inverted U-shape differential pressure gauge. The pipeline part of the setup includes three sections of parallel long straight pipes, which are used to determine respectively the local resistance coefficient, smooth straight pipe resistance coefficient and rough straight pipe resistance coefficient. A stainless steel pipe mounted with to-be-measured pipe fitting (i.e., gate valve) is used to measure the local resistance; a stainless steel pipe with smooth inner wall is also used to measure the straight pipe resistance of the smooth pipe; while the straight pipe resistance of the rough pipe is measured by using a galvanized pipe with rather rough inner wall. The flowing volume is measured by the turbine flowmeter, whose signal is transmitted on to the

Chapter 1 Typical Momentum Transfer Experiment

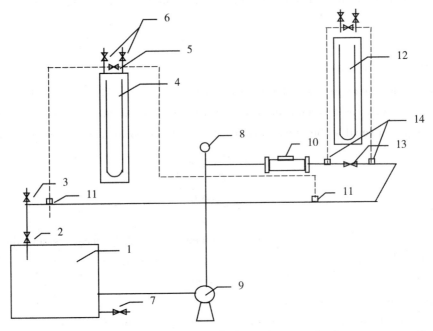

1: water tank; 2: control valve; 3: vent valve; 4: U-shape piezometer for the measurement of straight pipe resistance; 5: balance valve; 6: vent valve; 7: discharge valve; 8: thermometer; 9: pump; 10: turbine flowmeter; 11: straight pipe section pressure gauging hole; 12: U-shape piezometer for the measurement of local resistance; 13: gate valve; 14: local resistance pressure gauging hole.

Figure 1-11 Schematic Diagram of Fluid Flow Resistance Experiment

corresponding display instrument to show the rotation rate; the resistance of the pipeline and fittings is directly read from the inverted U-shape differential pressure gauge.

(3) Device Parameter

The device parameters are shown in Table 1-3. Since the material of the pipes differs by the batch, their diameters may be varied somewhat, as a result of which the inner diameters of the pipe given in Table 1-3 are the averaged number.

Table 1-3 **Device Parameter**

	Title	Material	Pipe Inner Diameter (mm)		Length of Measured Section (cm)
			Pipe No.	Pipe Inner Diameter	
Device	Local Resistance	Gate Valve	1A	20.0	95
	Smooth Pipe	Stainless Steel Pipe	1B	20.0	100
	Rough Pipe	Galvanized Iron Pipe	1C	21.0	100

1.2.4 Simulation Devices of the Experiment

The virtual simulation software interface for the determination of the fluid resistance in the pipe is shown in Figure 1-12. Wherein, regarding the smooth pipe: glass pipe diameter = 20 mm, length = 1.5 m, absolute roughness = 0.002 mm; regarding the rough pipe: galvanized iron pipe inner diameter = 20 mm, length = 1.5 m, absolute roughness = 0.2 mm; regarding the sudden expansion pipe: inner diameter of thin section = 20 mm, inner diameter of thick section = 40 mm; regarding the orifice flowmeter: orifice diameter = 12 mm, orifice flow coefficient = 0.62.

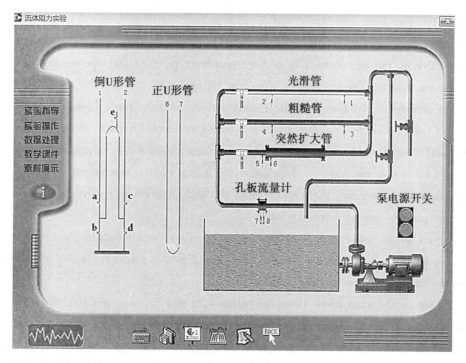

Figure 1-12 The Virtual Simulation Software Interface for the Determination of the Fluid Resistance in the Pipe

1.2.5 Experimental Procedure

(1) Preparation for the Experiment

① Clean the water tank, remove the debris from the bottom to protect the impeller of the pump and the turbine flowmeter. Close the blow-out valve at the bottom of the tank, fill in water up to about 15cm short of the upper edge of the tank to both provide enough water for the experiment and prevent splashing at the outlet pipe.

② Plug in the power supply of the control cabinet, turn on the master switch and the gauge power supply, and perform self-inspection of the instruments. Open the ball valve between the water tank and the pump, turn off the reflow valve of the pump and open the gate valve under the rotor flow gauge to the full. After the above steps, if the pump fails to lift water, it may be caused by the reversal of the impeller. Firstly, check any lack of phase, which usually can be judged from the indicator light. Secondly, check any reversal of phase. It is necessary to check the phase sequence of the power supply of the centrifugal pump motor, and adjust any two-line socket of the three fire lines.

(2) Selection of Pipeline for the Experiment

Select the pipeline for the purpose of the experiment, open the corresponding inlet valve, and keep the full flow for about 5~10 minutes at the maximum opening of the outlet valve.

(3) Air Expulsion

Firstly, perform the pressure drawing operation in the pipeline. It is necessary to open the impulse valve on the equalizing ring of the experiment pipeline and operate the inverted U-shape pipe as follows. The structure is shown in Figure 1-13.

① Expel the air bubbles from the system and the impulse pipe. Close the master outlet valve 9 of the pipeline to keep the system in a zero-flow, high-lift state. Close the intake valve 3, the outflow tap 5 and balance valve 4. Open the high-pressure-side valve 2 and low-pressure-side valve 1, so that the water of the experiment system can be discharged out of the system through the system pipeline, the impulse pipe, the high-pressure-side valve 2, the inverted U-shape pipe and the low-pressure-side valve 1.

② Suck in air through the glass pipe. After expelling the air bubbles, close the valves 1 and 2, open the balance valve 4, outflow tap 5 and intake valve 3, so that the glass pipe discharges all its water content and suck in air.

③ Balance the water level. Close the valves 4, 5 and 3, then open the valves 1 and 2 to let the water enter the glass pipe to the balance water level. At this time, the water outlet valve in the system is always closed. When there is no flow in the pipeline, the water level in the U-shape pipe is balanced, and the pressure difference gauge is in the stand-by state.

④ The differential pressure corresponding to the measured object at different flow rates is reflected as the difference between the left and right columns of the inverted U-shape pipe pressure difference gauge.

(4) Flow Regulation

Conduct the pipeline pressure difference test under different flow rates. Let the flow rate change from 0.8 to 4 m^3/h, advisably by 0.5m^3/h between experiments. Adjust the master outlet valve from large to small or vice versa. After every change to the flow rate, wait until the flow is stable to read the data. Record a total of 8~10 sets of experiment data. The main experiment parameters to be obtained are: flow rate Q, pressure difference of the measured

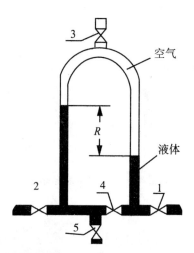

1: Low-pressure-side valve; 2: Hight-pressure-side Valve; 3: Intake Valve;
4: Balance Valve; 5: Water outlet valve

Figure 1-13 Inverted U-shape differential pressure gauge

section Δp, and the fluid temperature t.

(5) Conclusion of Experiment

After the experiment is finished, close the master outlet valve of the pipeline, then close the master pump switch and turn off the power supply to the control cabinet; close the inlet ball valve of the pipeline and the impulse valve on the corresponding pressure equalizing ring, and clean the experiment devices (if to be left idle for a long time, discharge the residual water from the pipeline using the emptying valve and the water in the water tank can also be discharged from the drain valve).

1.2.6 Procedure of Simulation Experiment

Step one: turn on the pump (see Figure 1-14). Since the centrifugal pump is installed below the water level, there is no need for priming. Directly press the green button of the power switch to turn on the centrifugal pump and start the experiment.

Step two: expel air from the pipe system and adjust the inverted U-shape differential pressure gauge (See Figure 1-15). Open all the valves in the pipe system, so that water flows in the 3 pipelines for a period of time until the air in the system is cleared off. Then click on the inverted U-shape pipe, and there will be animation of the adjustment of the inverted U-shape pipe. Finally close the valves and start the test operation.

Step three: Measure the data of the smooth pipe (See Figure 1-16).

(1) Establishing Flow in the Smooth Pipe

Chapter 1 Typical Momentum Transfer Experiment

Figure 1-14 Schematic Diagram of Turning on the Pump

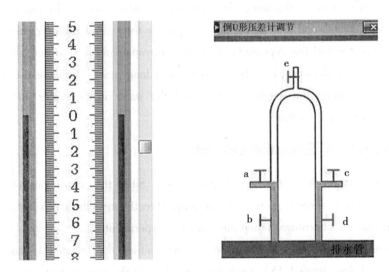

Figure 1-15 U-shape Differential Piezometer

After turning on the centrifugal pump and adjusting the U-shape differential pressure gauge, the flow can be established by successively adjusting opening of valve 1, valve 2 and valve 3 to over 0, as shown in the Figure 1-16. Close the ball valve of the rough pipe and

1.2 Determination of Fluid Resistance in Pipe

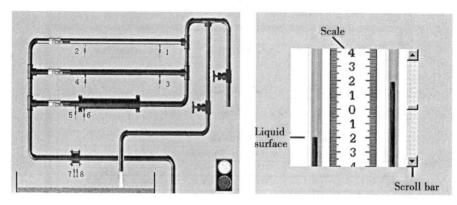

Figure 1-16 Smooth Pipe Test

sudden increasing pipe, open the ball valve of the smooth pipe to make the water flow only in the smooth pipe.

(2) Reading the Data

Click the left button of the mouse on the U-shape or inverted U-shape differential pressure gauge to see the picture as shown in Figure 1-16 (the red level is only used for demonstration, in practice, the device may be of any other color, for example, the mercury is silver white in color). Join the pressure intake of the inverted U-shape differential pressure gauge that of the pipe, and join the pressure intake of the U-shape differential pressure gauge with that of the orifice plate. Use the mouse to drag the scroll bar up or down to read the data. In the experiment, each pipe has an inverted U-shape pipe, and click continuously the inverted U-shape pipes in Figure 1-16 can switch between the three. The number above the inverted U-shape pipe indicates the pipe connected to it. Note: The data read is of the height difference between two liquid levels, unit: mm.

(3) Recording the Data

Click the left button of the mouse on the "Data Processing" button in the menu on the left side of the main screen to call out the data processing window (See Figure 1-17), click on the original data page and fill in the data read from the U-shape differential pressure gauge and the inverted U-shape differential pressure gauge in the U-shape differential pressure gauge column and the inverted U-shape differential pressure gauge column respectively following the standard database operation method.

Note: If the automatic recording function is used, when the "Automatic Record" key is clicked, the data will be automatically written with no need for manual typing.

(4) Recording Multiple Sets of Data

Adjust the opening of the valve to change the flow rate, repeat steps (2) and (3) above,

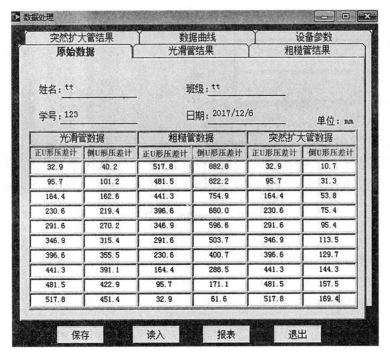

Figure 1-17　Data Processing Window

and for the purpose of experimental accuracy and plotting the regression curve, measure at least ten sets of data. Afterwards, enter the next step which is to measure the data of the rough pipe.

Step four: Measure the data of the rough pipe.

(1) Establishing Flow in the Rough Pipe

After completing the measurement and recording of the smooth pipe data, establish the flow in the rough pipe.

(2) Measuring and Recording the Data

Measure the data of the rough pipe in the same way as measuring the data of the smooth pipe. Repeat steps (2)(3)(4) in measuring the data of the smooth pipe. For the purpose of experimental accuracy and plotting the regression curve, measure at least ten sets of data. Afterwards, enter the next step which is to measure the data of the sudden increasing pipe.

Step five: Measure the data of the sudden increasing pipe.

(1) Establishing Flow in the Sudden Increasing Pipe

After completing the measurement and recording of the rough pipe data, establish the flow in the sudden increasing pipe.

(2) Measuring and Recording the Data of the Sudden Increasing Pipe

Measure the data of the sudden increasing pipe in the same way as measuring the data of the smooth pipe. Repeat steps (2)(3)(4) in measuring the data of the smooth pipe. For the purpose of experimental accuracy and plotting the regression curve, measure at least ten sets of data. Afterwards, proceed to data processing.

Attention:

① In order to approach the ideal smooth tube, a glass pipe is selected. However, in practice, glass pipes are rarely used in ordinary laboratories.

② For the better regression treatment of the data, measure as much data as possible, and try to spread the data in the whole flow range if possible.

③ Within the scope of the static flow, it is hard to ensure the regulating precision when using the valve button, please type in the degree of opening in the valve opening column (the valve opening should be less than 5).

④ For the sudden increasing pipe, simplification is made, where the resistance coefficient is deemed a fixed value and does not change with Re.

1.2.7 Data Recording and Processing

Fill the data obtained from the above experiments in Table 1-4:

Basic parameters of the straight pipe: Smooth pipe diameter _____ ; Rough pipe diameter _____ ; Local resistance pipe diameter _____ .

Table 1-4　　**Experiment Data Recording Table**

No.	Flow Rate (m^3/h)	Smooth Pipe mmH_2O			Rough Pipe mmH_2O			Local Resistance mmH_2O		
		Left	Right	Pressure Difference	Left	Right	Pressure Difference	Left	Right	Pressure Difference

1.2.8 Recording and Processing of Simulation Experiment

Step one: Record the original data (See Figure 1-18), note: Since the three sets of data are in the same format, do not confuse them.

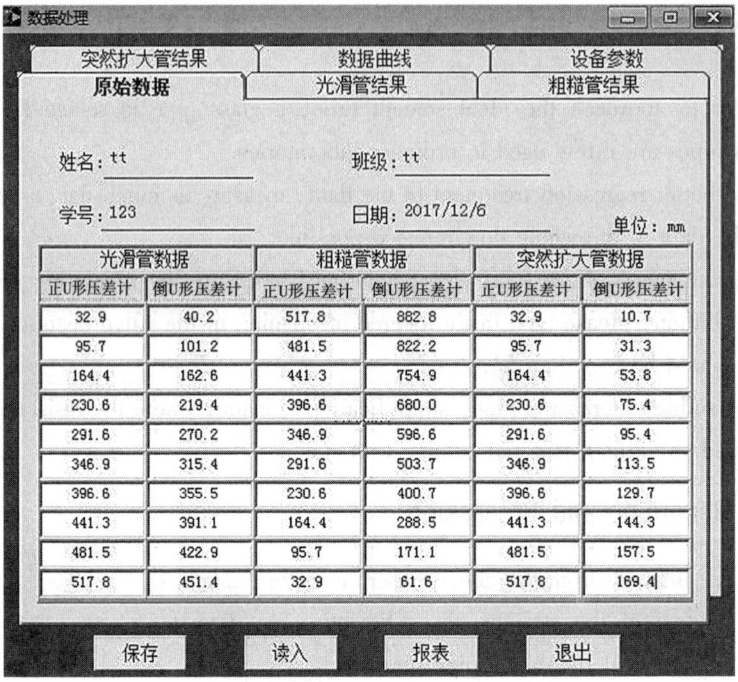

光滑管数据		粗糙管数据		突然扩大管数据	
正U形压差计	倒U形压差计	正U形压差计	倒U形压差计	正U形压差计	倒U形压差计
32.9	40.2	517.8	882.8	32.9	10.7
95.7	101.2	481.5	822.2	95.7	31.3
164.4	162.6	441.3	754.9	164.4	53.8
230.6	219.4	396.6	680.0	230.6	75.4
291.6	270.2	346.9	596.6	291.6	95.4
346.9	315.4	291.6	503.7	346.9	113.5
396.6	355.5	230.6	400.7	396.6	129.7
441.3	391.1	164.4	288.5	441.3	144.3
481.5	422.9	95.7	171.1	481.5	157.5
517.8	451.4	32.9	61.6	517.8	169.4

Figure 1-18 Record of the Original Data

Step two: Data calculation. After filling the data, if the "Automatic Calculation" function is not used, the equipment parameters required in the calculation can be obtained in the "Device Parameter" page of the data processing window. If the "Automatic Calculation" function is used, clicking the "Automatic Calculation" button on the corresponding results page will suffice. The data will be automatically calculated and filled in the database.

Step three: Plotting the curve. After the calculation is completed, the curve can be plotted automatically according to the data by clicking the "Automatic Plotting" button in the curve page as shown in Figure 1-19.

1.2.9 Discussion

① Does the outlet valve at the tail of the system have to be closed when the device is undergoing air expulsion? Why?

② How to check if the air in the pipeline has been completely expelled?

③ Can the $\lambda \sim Re$ relationship obtained with water as the medium be applied to other fluids? How to apply?

④ Can the $\lambda \sim Re$ data obtained at different water temperatures and with different devices

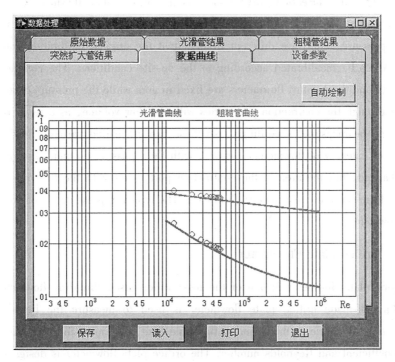

Figure 1-19 Curve Plotting Interface

(including pipes of different diameters) be correlated on the same curve?

⑤ If there is burr on the edge of the pressure gauging port or the orifice or the installation is not vertical, what is the effect on the static pressure measured?

1.3　Control Test of Flowmeter

1.3.1　Purpose of the Experiment

① To familiarize with the structure, performance and installation method of the orifice plate flowmeter and Venturi flowmeter.

② To master the capacity calibration method of the flowmeter.

③ To determine the relation between the orifice flow coefficient and Reynolds number of the orifice plate flowmeter and the Venturi flowmeter.

1.3.2　Principle of the Experiment

All non-standard flowmeters have to undergo flow calibration, flow calibration scale establishment (such as the rotator flowmeter), and offer orifice flow coefficient (such as the turbine flowmeter) and correction curve (such as the orifice plate flowmeter) determination before leaving factory. When in application, if operating conditions such as working medium, temperature, pressure, etc., are different from that at the original calibration condition, the flowmeter needs to be recalibrated according to the on-site conditions. The contraction port of the orifice plate and the Venturi flowmeters are fixed in area while the pressure drop of the fluid when flowing through the contraction port varies with the flow volume, as a result of which the flow volume can be measured accordingly, so these two are called variable head flowmeters. There is another category of flowmeters, where, with fluid passing through, the pressure drop remains unchanged, but the area of the contraction port varies with the flow rate. Therefore, they are called the variable cross-section flowmeter. The typical representative of this category is the rotator flowmeter.

1.3.2.1 Orifice Plate Flowmeter

The orifice plate flowmeter is one of the most widely used throttling flowmeters. In this experiment, a self-made orifice plate flowmeter is used to measure the liquid flow rate, calibrated by the volumetric method, and at the same time, determine the relationship between orifice flow coefficient and Reynolds number. The orifice plate flowmeter is designed according to the principle of the mutual transformation of the kinetic energy and the potential energy of fluid. When the fluid passes through a sharp orifice, the flow rate will increase, resulting in pressure difference between the front and back of the orifice plate. It can be displayed on the pressure difference gauge or differential pressure transmitter through an impulse pipe. Its basic structure is shown in Figure 1-20.

If the pipe diameter is d_1, the sharp orifice diameter is d_0, the diameter of the reduced flow formed when fluid flows through the orifice plate is d_2, the fluid density is ρ, then according to Bernoulli equation, the following equation can be established at the interface 1 and 2:

$$\frac{u_2^2 - u_1^2}{2} = \frac{p_1 - p_2}{\rho} = \frac{\Delta p}{\rho} \tag{1-20}$$

Or
$$\sqrt{u_2^2 - u_1^2} = \sqrt{2\Delta p/\rho} \tag{1-21}$$

Since the position of the reducing flow varies with the flow rate, the cross-sectional area A_2 is difficult to know, and the area of the orifice A_0 is known, the flow rate u_0 at the orifice is used to replace the u_2 in the equation (1-21), and considering the error brought by this replacement and the energy loss caused by the actual local resistance of the fluid, it is necessary to be corrected with the coefficient C. Equation (1-21) is rewritten to:

$$\sqrt{u_2^2 - u_1^2} = C\sqrt{2\Delta p/\rho} \tag{1-22}$$

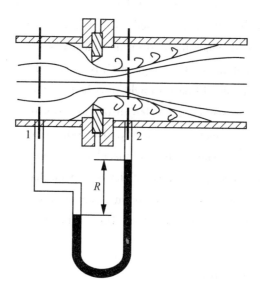

Figure 1-20 Orifice Plate Flowmeter

For incompressible fluids, it can be seen from the continuity equation that $u_1 = \dfrac{A_0}{A_1} u_0$, bringing which to equation (1-22), we have:

$$u_0 = \frac{C\sqrt{2\Delta p/\rho}}{\sqrt{1 - \left(\dfrac{A_0}{A_1}\right)^2}} \tag{1-23}$$

Make
$$C_0 = \frac{C}{\sqrt{1 - \left(\dfrac{A_0}{A_1}\right)^2}} \tag{1-24}$$

Then equation (1-23) can be simplified to:

$$u_0 = C_0 \sqrt{2\Delta p/\rho} \tag{1-25}$$

The volume flow of fluid can be calculated according to u_0 and A_0.

$$V = u_0 A_0 = C_0 A_0 \sqrt{2\Delta p/\rho} \tag{1-26}$$

Or
$$V = u_0 A_0 = C_0 A_0 \sqrt{2gR(\rho_i - \rho)/\rho} \tag{1-27}$$

Wherein: V: the volume flow of the fluid (m^3/s); R: the reading of the U-shape differential pressure gauge (m); ρ_i: the density of the indicator liquid in the differential pressure gauge (kg/m^3); C_0: the orifice flow coefficient, dimensionless.

C_0 is determined by the shape of the sharp orifice, the position of the pressure gauging port, the ratio between the diameters of the orifice and the pipe, and the Reynolds number Re, the specific value of which has to be determined by experiment. When the ratio of the diameters of the orifice to the diameter of the pipe is fixed, if Re is over a certain value, C_0 will approach to be constant. The flowmeters generally used in the industry is defined to be used in the flow conditions where C_0 is a fixed value. The range of C_0 is usually $0.6 \sim 0.7$. When installing the orifice plate flowmeter, there should be a straight pipe both in the upper and lower reaches of the flowmeter to act as stable sections, the length of the upstream section should be at least $10d_1$, and that of the downstream at least $5d_2$. The orifice plate flowmeter is simple in construction and convenient in manufacture and installation, however, its main disadvantage is the large loss of mechanical energy. Because of the loss of mechanical energy, after the fluid recovers its flow rate at the downstream, the pressure cannot be restored to the value in front of the orifice plate, which is called permanent loss. The smaller the value of d_0/d_1, the greater the permanent loss is.

1.3.1.2 Venturi Flowmeter

The main drawback of the orifice plate flowmeter is its great mechanical energy loss. In order to overcome this shortcoming, a converging-diverging tube can be used. As shown in Figure 1-21, when the fluid flows through such a conical tube, there will be no boundary layer separation and whirlpool, which greatly reduces the loss of mechanical energy. This tube is called the Venturi tube. The contracting cone angle of the Venturi tube is usually between $15° \sim 25°$, the expanding cone angle needs to be smaller, generally between $5° \sim 7°$, to let the flow rate change more smoothly, because the mechanical energy loss mainly occurs in the sudden increasing section.

The working principle of the Venturi flowmeter is exactly the same as that of the orifice plate flowmeter, but the permanent loss is much smaller. The flow rate and volume can still be calculated as per equations (1-25) and (1-26), wherein u_0 still represents the flow rate at the minimum cross-section area (called the Venturi throat). The orifice flow coefficient C_0 of the Venturi tube is about $0.98 \sim 0.99$. The mechanical energy loss is:

$$w_f = 0.1 u_0^2 \tag{1-27}$$

The drawback of the Venturi flowmeter is that its processing is more complex than the orifice plate flowmeter, thus its cost is higher, besides, its installation needs to take up a certain length of the pipe. However, with its small permanent loss, it is particularly suitable for the transportation of low pressure gas.

1.3.3 Experimental Devices

The experimental devices are shown in Figure 1-22. The main part of the setup includes the circulating water pump, flowmeter, U-shape differential pressure gauge, thermometer and

1.3 Control Test of Flowmeter

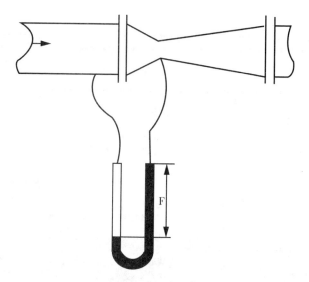

Figure 1-21　Venturi Flowmeter

water tank, etc. The main pipe of the experiment is a 1-inch stainless steel pipe (25mm of inner diameter).

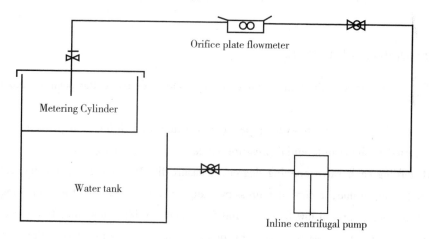

Figure 1-22　Schematic Diagram of Control Test of Flowmeter

1.3.4　Simulation Devices of the Experiment

The virtual simulation interface of the flowmeter control test is shown in Figure 1-23, device parameters: area of the measuring tank: $1m^2$; the diameter of the pipe: 30mm; the diameter of the orifice: 20mm.

Chapter 1 Typical Momentum Transfer Experiment

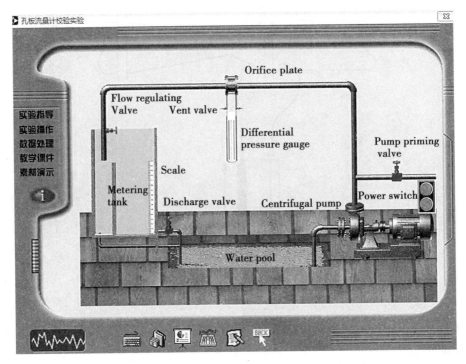

Figure 1-23 Virtual Simulation Interface of Flowmeter Check Test

1.3.5 Experimental Procedure

① To be familiar with the experimental devices and understand the position and function of each valve.

② To expel air from the relevant pipe, impulse tube and differential pressure gauge, so that the inverted U-shape differential pressure gauge is in the working state.

③ Regarding each valve opening degree, measure the flow volume using the volumetric method, at the meantime, record the pressure gauge readings, and in the ascending order, measure about 8~9 points in low flow rate and 5~6 points in high flow rate. In order to ensure calibration accuracy, it is best to repeat the process, this time in the descending order, and then take the average value of the two.

④ When measuring the flow rate, it is necessary to ensure that the liquid level difference of the metering bucket is not less than 100mm or the measuring time is not less than 40s in each measurement.

⑤ The main calculation process is as follows:

a. Calculate the flow volume V (m^3/h) using the volumetric method (with stopwatch and metering tube).

b. According to $u = \dfrac{4V}{\pi d^2}$, calculate the value of u.

c. Read the height difference R of the differential pressure meter corresponding to the flow volume V (regulated by the gate valve opening degree), then calculate the value of C_0 according to $u_0 = C_0 \sqrt{2\Delta p/\rho}$ and $\Delta p = \rho g R$.

d. Calculate the Reynolds number according to $Re = \dfrac{du\rho}{\mu}$, wherein d has the value of the corresponding d_0.

e. Plot the C_0-Re diagrams of the orifice plate flowmeter and the Venturi flowmeter on the coordinate paper.

1.3.6 Procedure of Simulation Experiment

Step one: Prime the pump (See Figure 1-24).

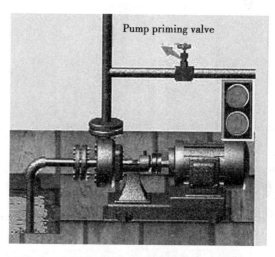

Figure 1-24 Pump Priming

Since the installation height of the centrifugal pump is above the liquid level, it is necessary to prime the centrifugal pump before starting it. Since the focus of this experiment is on the flowmeter instead of the centrifugal pump, the priming process is simplified. As shown in Figure 1-24, as long as the priming valve opening degree is greater than 0, wait for more than 10 seconds, and then close it, the system will deem the priming operation has been completed.

Step two: Turn on the pump (See Figure 1-25). After completing the pump priming, click the green button of the power switch to connect the power supply, when the centrifugal pump can be turned on and start working.

Step three: After starting the centrifugal pump, adjust the main regulating valve to be opened at 100, as shown in Figure 1-26.

Chapter 1 Typical Momentum Transfer Experiment

Figure 1-25 Turning on the Pump

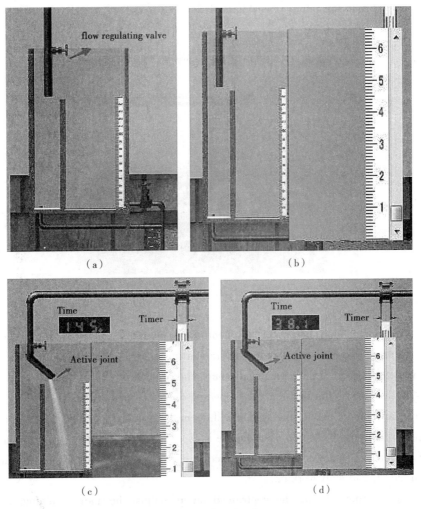

Figure 1-26 Flow Regulation

Step four: Reading the Data.

Click the left button of the mouse on the scale to call out the reading screen of the scale. First record the initial height of the liquid level. Click the right button of the mouse to close the scale screen. Then click the left button of the mouse on the mobile connector to divert the water to the metering tank, and it can be seen that the liquid level starts rising and the timer automatically starts keeping time. When the liquid level rises to a certain height, click the left button of the mouse on the mobile connector to divert the liquid to the discharge part, and at the same time the timer will stop automatically. At this time, the height of the liquid level and the reading of the timer will be recorded. Click the left button of the mouse on the differential pressure gauge and drag the scroll bar with the mouse to read its data.

Step five: Record the data (See Figure 1-27).

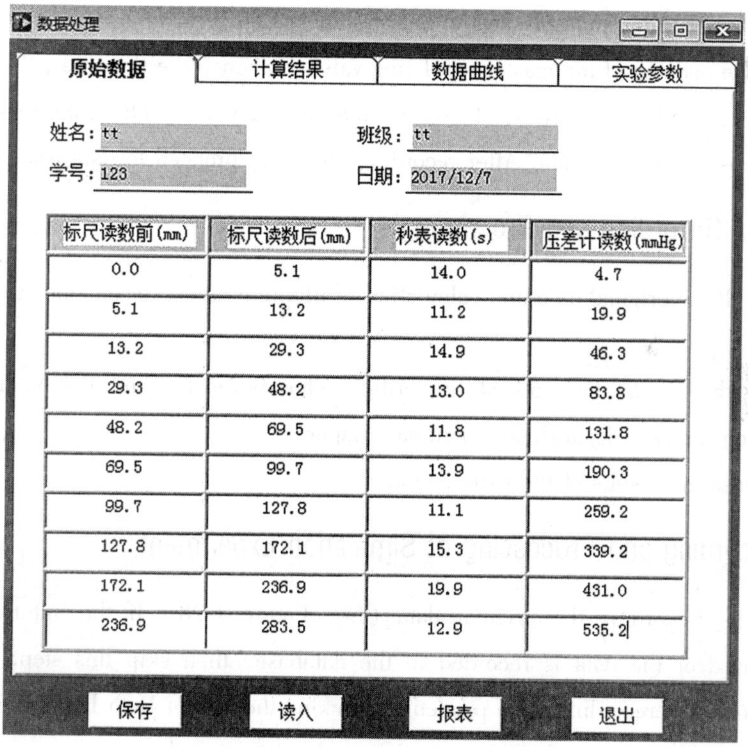

标尺读数前(mm)	标尺读数后(mm)	秒表读数(s)	压差计读数(mmHg)
0.0	5.1	14.0	4.7
5.1	13.2	11.2	19.9
13.2	29.3	14.9	46.3
29.3	48.2	13.0	83.8
48.2	69.5	11.8	131.8
69.5	99.7	13.9	190.3
99.7	127.8	11.1	259.2
127.8	172.1	15.3	339.2
172.1	236.9	19.9	431.0
236.9	283.5	12.9	535.2

Figure 1-27 Data Recording

Click the left button of the mouse on the "Data Processing" button in the menu on the left side of the main screen to call out the data processing window, click on the original data page and fill in the data read from the U-shape differential pressure gauge and the inverted U-shape differential pressure gauge in the U-shape differential pressure gauge column and the inverted U-shape differential pressure gauge column respectively following the standard database operation method. Or the data can be recorded in the data recording table printed by clicking the "Print Data Recording Table" key.

Attention: If the automatic recording function is used, when the "automatic record" key is clicked, the data will be automatically written with no need for manual typing. In order to better show the change of the orifice flow coefficient C_0 with Re when Re is rather small, the flow volume in the experiment is set very low to obtain a smaller Re. In addition, the general flowmeter calibration control test takes the average of multiple measurements with the orifice flow coefficient almost unchanging so as to obtain C_0, instead of allowing C_0 to change with Re. Therefore, if the data and calculation are recorded manually, the error would be significant. As a result, better results can be obtained by automatic calculation.

Step six: Recording multiple sets of data.

Adjust the opening of the master regulating valve to change the flow rate, then repeat steps 4 and 5 above, and for the purpose of experimental accuracy and plotting the regression curve, measure at least ten sets of data. After recording the data, proceed to data processing.

1.3.7 Data Recording and Processing

① List all the original data and calculation results in a table, accompanied with numerical examples.

② Plot the C_0-Re diagrams of the orifice plate flowmeter and the Venturi flowmeter respectively on the semi-logarithmic coordinate paper.

③ Discuss the results of the experiment.

1.3.8 Recording and Processing of Simulation Experiment

Step one: Recording the original data (See Figure 1-28). If the automatic recording function is used or the data is recorded in the database, then skip this step. If the data is recorded in the data recording table printed by clicking the "Print Data Recording Table" key, please fill the data in the database by referring to the data record.

Step two: Data calculation (See Figure 1-29). If the automatic calculation function is used, clicking the "Automatic Calculation" button on the corresponding results page will suffice, as shown in Figure 1-29. The data will be automatically calculated and filled in the database.

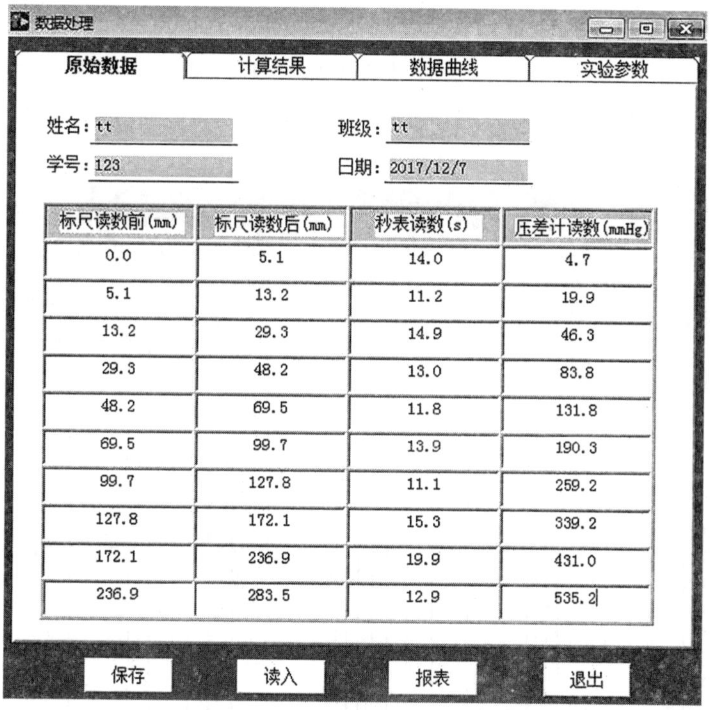

1.3 Control Test of Flowmeter

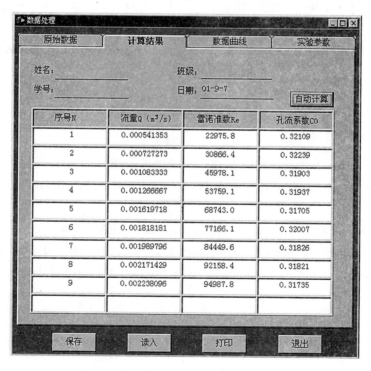

Figure 1-28 Record of the Original Data

Figure 1-29 Result of Automatic Calculation

Step three: Plotting the curve. After the calculation is completed, the curve can be plotted automatically according to the data by clicking the "Automatic Plotting" button in the curve page as shown in Figure 1-30.

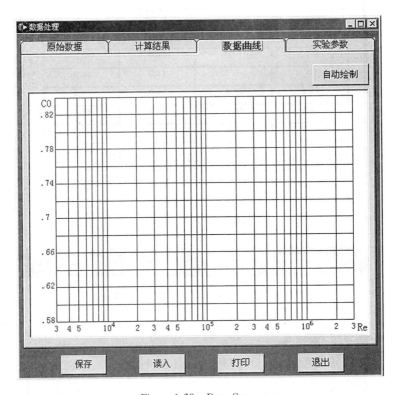

Figure 1-30 Data Curve

1.3.9 Discussion

① What are the factors relating to the orifice flow coefficient?

② What are the issues to note when installing the orifice plate flowmeter and the Venturi flowmeter?

③ How to check if the air in the system has been completely expelled?

④ In the experiment, the correction curve of ΔR-V can be obtained directly, which, after conversion can also derive the C_0-Re curve. What are the advantages of the two methods?

Chapter 2　Typical Heat Transfer Experiment

2.1　"Air-water" Heat Transfer Experiment

2.1.1　Purpose of the Experiment

① To master the method of determining the heat transfer coefficient K.
② To learn the operation method of the heat exchanger.

2.1.2　Principle of the Experiment

The heat exchanger is a type of common heat transfer equipment used in industrial production. Through the heat transfer wall, the hot fluid passes its heat to the cold fluid to meet the requirements of the production process. The parameters affecting the heat transfer volume of the heat exchanger include heat transfer area, average temperature difference and heat transfer coefficient. In order to properly select or design the heat exchanger, the performance of the heat exchanger should be fully understood. In addition to referring to literature, the performance measurement test of the heat exchanger is an important way for this purpose. The heat transfer coefficient is an important index to measure the performance of heat exchanger. In order to improve the utilization rate of energy, how to enhance the heat transfer coefficient of the heat exchanger and strengthen the heat transfer process is a common problem encountered in the production practice. The tubular heat exchanger is an inter-wall heat transfer device whose heat transfer process between the cold and hot liquids includes the three sub-processes of heat transfer between hot fluid and wall surface, the heat conduction of the solid wall surface and the heat convection of the wall surface to the cold liquid (see Figure 2-1).

The heat transfer rate is calculated as per equation (2-1):

$$Q = KA \frac{(T_{outlet} - t_{inlet}) - (T_{inlet} - t_{outlet})}{\ln \dfrac{T_{outlet} - t_{inlet}}{T_{inlet} - t_{outlet}}} = c_{pg} q_{vg} \rho (T_{inlet} - T_{outlet}) \tag{2-1}$$

Wherein: K: heat transfer coefficient ($W/m^2 \cdot °C$); A: heat transfer area (m^2); T_{inlet}: the entering temperature of the hot fluid (°C); T_{outlet}: the exiting temperature of the hot fluid

Chapter 2　Typical Heat Transfer Experiment

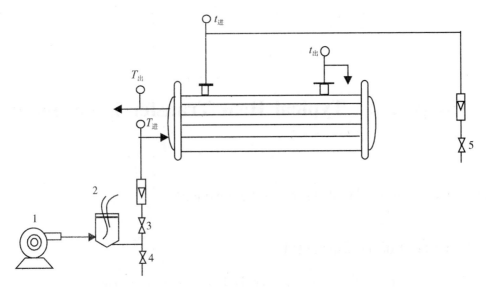

Figure 2-1　Schematic Diagram of the Gas-liquid Heat Transfer Experiment

(℃); t_{inlet}: the entering temperature of the cold fluid (℃); t_{outlet}: the exiting temperature of the cold fluid (℃); c_{pg}: specific heat at constant pressure of the hot fluid (J/kg · ℃); q_{vg}: flow volume of the cold fluid (m³/s); ρ: density of the hot fluid density (kg/m³).

2.1.3　Experimental Devices

In this experiment, the cold fluid is water and the hot fluid is air. Flowing out of its source, the cold fluid is first measured of its flow rate by a rotor flowmeter and its entering temperature by a thermometer before entering the shell pass of the heat exchanger. After the exchange, it has its exiting temperature measured at the outlet; flowing out of the air source, the hot fluid first has its flow rate measured by a rotor flowmeter and then reaches the pipe pass of the heat exchanger after being heated to 120℃. Its entering temperature is measured at the inlet and exiting temperature at the outlet. The experimental devices are as shown in Figure 2-2.

2.1.4　Experimental Procedure

① Open the source of the cold fluid, and adjust its flow rate using the regulating valve 5.

② Open the air source 1 and the valve 4, adjust the air flow rate using the regulating valve 3, switch on the power, set the temperature to be between 100 ~ 120℃ using the intelligent temperature regulator AI-708.

③ Keep the flow rate of the hot and cold fluids constant, when the entering temperature of the hot air is basically unchanging in a certain period of time (like 10 minutes), record the relevant data.

④ When measuring the heat transfer coefficient K, while maintaining the flow rate of the

Figure 2-2 Schematic Diagram of Heat Transfer Experiment Setup

cold fluid constant, change the flow rate of the hot air a number of times according to the requirement of the experiment steps.

⑤ At the end of the experiment, turn off the heating power, wait until the temperature of the hot air falls below 50℃ to close the regulating valve of both the cold and hot fluids and close the sources of the cold and hot fluids.

2.1.5 Data Recording and Outcome Processing

Please record the experiment data in table 2-1. Heat transfer area A _____ m^2; Flow volume of cold fluid _____ L/h.

Table 2-1   **Experiment Record Table**

No. \ Parameter	T_{inlet}(℃)	T_{outlet}(℃)	t_{inlet}(℃)	t_{outlet}(℃)	q_{vg}(m^3/s)	K(W/(m^2·℃))
1						
2						
3						
4						
5						
6						
7						

2.1.6 Devices and Process of the Virtual Simulation Experiment

The experiment setup is shown in Figure 2-3, where the cold water enters the inner tube of the heat exchanger via a pump and a U-shape differential pressure gauge, and exchanges heat with the steams in the sleeve annulus. The flow volume of the cold water can be adjusted by the flow control valve. The steam rises from the steam generator into the sleeve annulus and exchanges heat with the cold water in the inner tube. The vent valve is used to discharge non-condensable gases. A certain length of stabilization section is set before the copper tube, which is to eliminate the end effect. The two ends of the copper tube are connected to the pipe with a plastic hose to eliminate the thermal stress. In this experiment, the cold water flows in the inner tube and the steam in the annulus (glass tube). The entering and exiting temperature of the water and the temperature of the tubular wall are respectively measured using the platinum resistance. The platinum resistance used to measure the entering and exiting temperature of the water should be placed at the center of the inlet and outlet. The platinum resistance used to measure the temperature of the tubular wall temperature is fixed at both ends of the outer wall of the inner tube with thermal insulation adhensive. The pressure difference of the orifice plate flowmeter is measured using a U-shape differential pressure gauge. In this experiment, the steam generator is made of stainless steel and installed with a glass liquid level meter. The thermal power of the generator is 1.5 kW.

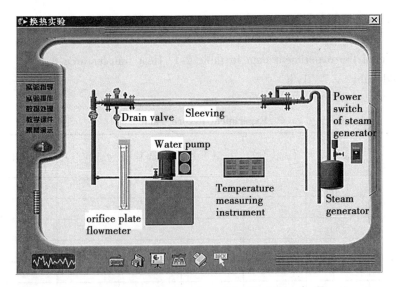

Figure 2-3 Virtual Simulation Interface of the Heat Transfer Experiment

2.1 "Air-water" Heat Transfer Experiment

2.1.7 Operation of Heat Transfer Virtual Simulation Experiment

Step one: Click the green button of the power switch to start the pump which will supply water for the tube pass of the heat exchanger (see Figure 2-4).

Figure 2-4　Start Position of Power Supply

Step two: Open the inlet valve, after opening the pump, adjust the inlet valve to slightly opening, when there would be water flowing in the tube pass of the heat exchanger (see Figure 2-5).

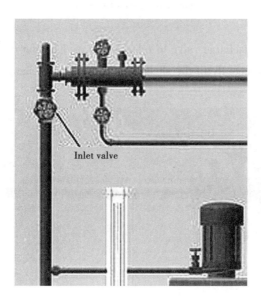

Figure 2-5　Switch of Inlet Valve

Step three: Start the steam generator, whose switch is on the right side of the **steam generator**. Click the left button of the mouse on the switch, when the steam generator will **start heating** and supply steam to the shell pass of the heat exchanger (see Figure 2-6).

Figure 2-6 Steam Generator

Step four: Open the discharge valve to release the residual non-condensate gases, so that **the steam** can flow unobstructedly in the shell pass of the heat exchanger (see Figure 2-7).

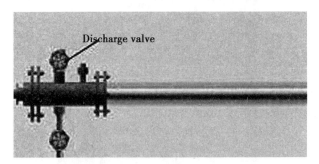

Figure 2-7 Discharge Valve

Step five: Read the flow rate of water. Click the differential pressure gauge of the **orifice plate flowmeter** in the picture to call out the reading screen. Read the differential **pressure**

gauge. The flow rate of the cold water can be obtained after calculation (see Figure 2-8).

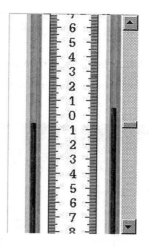

Figure 2-8　Flow Rate of Water

Step six: Read the temperature. Click on the heat exchange tube or the thermometer to call out the temperature reading screen. Read the temperature values at different positions. Among these, the temperature of the temperature nodes 1~9 is used to observe the distribution of the temperature and will not be used in data processing. The temperature at the inlet and outlet of the steam and water must be recorded. Click "Automatic Recording" button and the experimental data will be automatically recorded by the computer. Click the "Exit" button to close the temperature reading screen (see Figure 2-9).

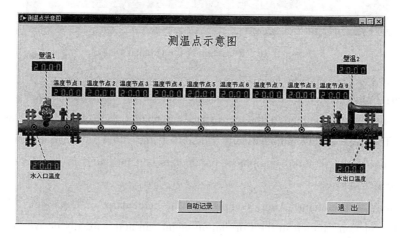

Figure 2-9　Temperature Reading

Step seven: Record multiple sets of data, change the opening of the inlet valve, repeat the above steps, and read 8 to 10 sets of data. After the conclusion of the experiment, first switch off the steam generator and then close the inlet valve.

2.1.8 Data Processing of the Virtual Simulation Experiment

Step one: Record the original data. The original data page is as shown in the Figure 2-10, through which the original data can be filled in and edited in data processing.

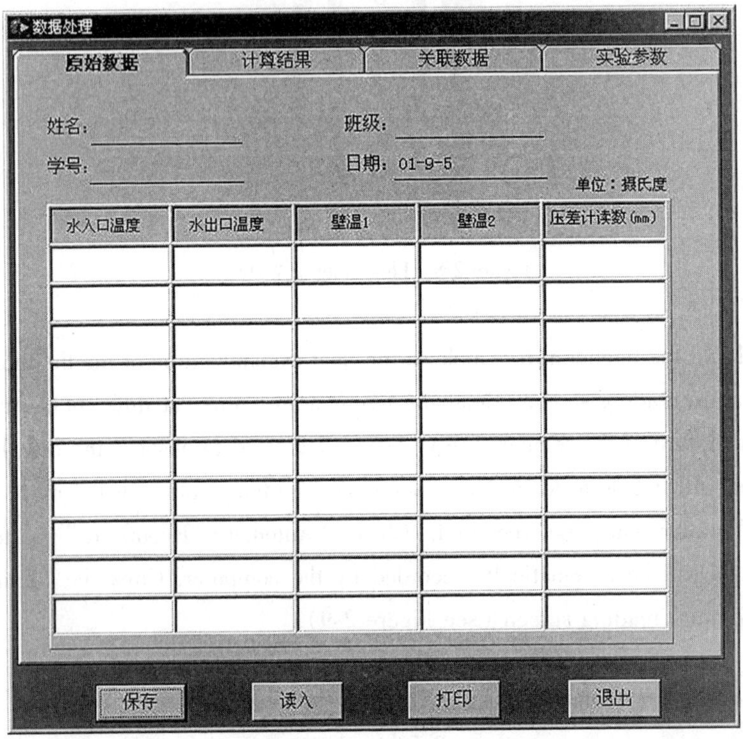

Figure 2-10 Recording of the Original Data

Step two: Data calculation. If the "Automatic Calculation" function is used, clicking the "Automatic Calculation" button on the corresponding results page will suffice, as shown in Figure 2-10. The data will be automatically calculated and filled in the database. If manual calculation is used, the required equipment parameters can be found in the device parameter page (see Figure 2-11).

Step three: Correlation. After completing the automatic calculation, clicking the "Automatic Correlation" button in the Correlation menu can automatically derive the correlative equation of the data (that is, giving the values in the place of the 0.000 and 0.00 in the Figure 2-12).

2.1 "Air-water" Heat Transfer Experiment

Figure 2-11 Calculation Result Table

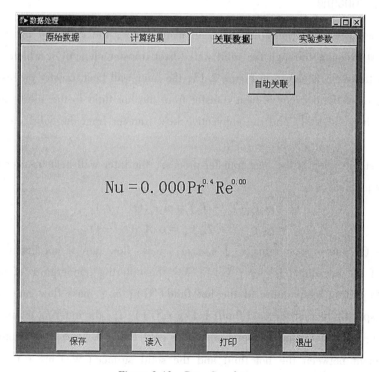

Figure 2-12 Data Correlation

2.1.9 Questions to Consider

How would the heat transfer rate Q and the heat transfer coefficient K change when the flow rate of the hot fluid is changed while the flow rate of the cold fluid is kept the same?

2.2 "Water-vapor" Heat Transfer Coefficient Experiment

2.2.1 Purpose of the Experiment

① To understand the inter-wall heat transfer elements and master the experimental method for determining the thermal transfer coefficient;

② To observe the condensation of water vapor on the outer wall of the horizontal tube and measure the forced convection heat transfer coefficient of water in a circular straight pipe;

③ To understand the factors affecting the heat transfer coefficient and the ways to strengthen heat transfer;

④ To understand the thermal resistance temperature measurement method, and the turbine flowmeter flow rate measurement method, and learn to use the frequency converter.

2.2.2 Basic Principle

In the process of industrial production, in many cases, the cold and hot fluid systems would have heat exchange through the solid wall (heat transfer element), which is called the inter-wall heat transfer. As shown in Figure 2-13, the inter-wall heat transfer process consists of the sub-processes of the convective heat transfer from the hot fluid to the solid wall, the heat conduction on the solid wall, and the convective heat transfer from the solid wall to the cold fluid.

After the stabilization of the heat transfer process, the inter-wall heat transfer element has the following equation:

$$Q = m_1 c_{p1} (T_1 - T_2) = m_2 c_{p2} (t_2 - t_1)$$
$$= \alpha_1 A_1 (T - T_W)_M = \alpha_2 A_2 (t_W - t)_m \quad (2-2)$$

Wherein: Q: rate of heat transfer (J/s); m_1: mass flow rate of hot fluid (kg/s); c_{p1}: specific heat of the hot fluid (J/(kg·℃)); T_1: the entering temperature of the hot fluid (℃); T_2: the exiting temperature of the hot fluid (℃); m_2: mass flow rate of cold fluid (kg/s); c_{p2}: specific heat of the cold fluid (J/(kg·℃)); t_1: the entering temperature of the cold fluid (℃); t_2: the exiting temperature of the cold fluid (℃); α_1: the convective heat transfer coefficient between the hot fluid and the solid surface (W/(m²·℃)); A_1: the convective heat transfer area on the side of the hot fluid (m²); $(T - T_W)_m$: logarithmic mean

2.2 "Water-vapor" Heat Transfer Coefficient Experiment

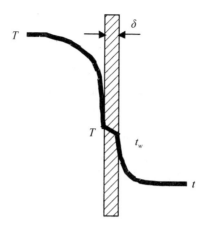

Figure 2-13 Schematic Diagram of the Inter-Wall Heat Transfet Process

of the temperature difference between the hot fluid and the solid wall(℃); α_2: the convective heat transfer coefficient between the cold fluid and the solid surface(W/(m² · ℃)); A_2: the convective heat transfer area on the side of the cold fluid(m²); $(t_W - t)_m$: logarithmic mean of the temperature difference between the solid wall and the cold fluid(℃);

The logarithmic mean of the temperature difference between the hot fluid and the solid wall can be calculated as per equation (2-3):

$$(T - T_W)_m = \frac{(T_1 - T_{W1}) - (T_2 - T_{W2})}{\ln \dfrac{T_1 - T_{W1}}{T_2 - T_{W2}}} \qquad (2-3)$$

Wherein, T_{W1}: the wall temperature by the side of the hot fluid at the inlet of the hot fluid (℃); T_{W2}: the wall temperature by the side of the hot fluid at the outlet of the hot fluid(℃).

The logarithmic mean of the temperature difference between the solid wall and the cold fluid can be calculated as per equation (2-4):

$$(t_W - t)_m = \frac{(t_{W1} - t_1) - (t_{W2} - t_2)}{\ln \dfrac{t_{W1} - t_1}{t_{W2} - t_2}} \qquad (2-4)$$

Wherein, t_{W1}: the wall temperature by the side of the cold fluid at the inlet of the cold fluid (℃); t_{W2}: the wall temperature by the side of the cold fluid at the outlet of the cold fluid (℃).

In the tubular heat exchanger, the steam flows in the heat exchange bucket and the water flows inside the copper tube; the steam would condense on the surface of the copper tube and heat up the water. After the heat transfer process is stable, there is equation (2-5):

$$V\rho C_P(t_2 - t_1) = \alpha_2 A_2 (t_W - t)_m \qquad (2-5)$$

Wherein, V: flow volume of the cold fluid(m^3/s); ρ: density of the cold fluid(kg/m^3); C_P: specific heat of the cold fluid($J/(kg \cdot ℃)$); t_1, t_2: the entering and exiting temperature of the cold fluid(℃); α_2: convective heat transfer coefficient of the cold fluid to the inner wall of the inner tube($W/(m^2 \cdot ℃)$); A_2: the heat transfer area on the inner wall of the inner tube (m^2); $(t_W-t)_m$: the logarithmic mean of the temperature difference between the inner wall and the fluid, available by reference to equation (2-3), (℃); when the material of the inner tube has good thermal conductivity, that is, when the λ value is large, and the wall is very thin, it can be deemed that $T_{W1}=t_{W1}$, $T_{W2}=t_{W2}$, that is the measurement of the wall temperature at the point. From equation (2-1), we have:

$$\alpha_2 = \frac{V\rho C_P(t_2 - t_1)}{A_2 (t_W - t)_m} \qquad (2-6)$$

If the value of V, t_1 and t_2 of the heated fluid, the heat exchange area A_2 of the inner tube, and the wall temperature t_{W1} and t_{W2} can be measured, we can calculate empirically the convective heat transfer coefficient α_2 of the cold fluid in the tube by equation (2-6).

For the fluid engaged in forced turbulent convective heat transfer in a circular straight pipe, the experiential formula of the heat transfer number is:

$$N_u = 0.023\ Re^{0.8} P_r^n \qquad (2-7)$$

Wherein, N_u: the Nusselt number, $N_u = \frac{\alpha d}{\lambda}$, dimensionless; Re: the Reynolds number, $Re = \frac{du\rho}{\mu}$, dimensionless; Pr: the Plante number, $P_r = \frac{c_p\mu}{\lambda}$, dimensionless; the applicable scope of equation(2-7) is: $Re = 1.0 \times 10^4 \sim 1.2 \times 10^5$, $P_r = 0.7 \sim 120$, ratio of pipe length to pipe inner diameter $\frac{L}{d} \geq 60$. When the fluid is heated, $n=0.4$, when the fluid is cooled, $n=0.3$. α: the convective heat transfer coefficient of the fluid to the solid wall($W/(m^2 \cdot ℃)$); d: inner diameter of the heat exchanger tube(m); λ: thermal conductivity coefficient of the fluid ($W/(m \cdot ℃)$); u: the average rate of the flow of the fluid in the tube (m/s); ρ: fluid density(kg/m^3); μ: the viscosity of the fluid($Pa \cdot s$); c_p: the specific heat of the fluid($J/(kg \cdot ℃)$). Therefore, the curve can be plotted after the fitting of the data points obtained in the experiment, which can be compared with the curve of the experiential formula to verify the result of the experiment.

2.2.3 Device and Process

The setup of the experiment consists of a steam generator, a glass rotor flowmeter, a tube heat exchanger, a temperature sensor and a temperature display, etc (see Figure 2-14). the parameters of the devices: The size of the copper tube: 12×2 mm, that is, the inner diameter is 8 mm, the length is 1m; the maximum flow rate of the cold fluid is 400 L/h, and the lower measuring limit of the turbine flowmeter is 40 L/h. The water-steam system: The steam from the steam generator enters the tube heat exchanger to have heat exchange with the water from the water tank and the condensed water will be discharged into the drain through the pipeline. The cold water enters the inner tube of the heat exchanger (copper tube) after being pressurized by the turbocharger pump and the rotor flowmeter. The water flow can be regulated and adjusted by a valve, and after the heat exchange the water will be discharged into the sewer. Note that the cold fluid is not reused in this experiment, and water should be continuously supplied to the water tank during the experiment.

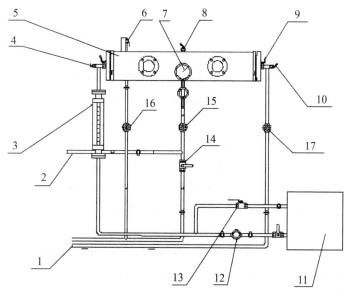

1: incoming pipes; 2: steam entrance; 3: glass rotor flowmeter; 4: cold fluid entering temperature;
5: tube heat exchanger; 6: inert gas; 7: steam inlet pressure gauge; 8: steam temperature;
9: steam exiting temperature; 10: cold fluid exiting temperature; 11: water tank; 12: water pump;
13: bypass valve; 14, 16: condensed water drainage valves; 15: steam inlet regulating valve;
17: fluid flow regulating valve

Figure 2-14　The Setup of the "Water-Vapor" Heat Transfer Coefficient Experiment

2.2.4　Experimental Procedure

(1) Manual Operation

① Check if the gauges, water pump, steam generator and temperature measuring points are working normally and check if the incoming steam regulating valve is closed.

② Turn on the master power and the instrument power (the steam generator and the master steam valve should be opened by the teacher).

③ Start the water pump.

④ Adjust the opening of the manual regulating valve to full opening to maximize the water flow rate.

⑤ Remove all the condensing water originally deposited in the steam pipeline (the way to do this: close the incoming steam regulating valve and open the condensate discharge valve of the steam pipe).

⑥ After the water is completed discharged, close the condensate discharge valve of the steam pipe, open the incoming steam regulating valve, so that the steam flows slowly into the heat exchanger annulus to heat up the tube heat exchanger; then open the condensate discharge valve of the heat exchanger (the condensate discharge valve should not be opened too much to avoid steam leakage) to let the condensed water in the annulus flow continuously to the trench.

⑦ Carefully adjust the opening of the incoming steam regulating valve, so that the steam pressure remains stable at about 0.1MPa (the target value of pressure can be reached by fine-tuning the inert gas emptying valve) to ensure the operations in the experiment are carried out under the condition of constant pressure. Then according to the test requirements, adjust the opening degree of the manual regulating valve gradually from large to small to determine six reasonable experimental points. After the flow and heat exchange are stable, read respectively the flow volume of the cold fluid, the entering and exiting temperature of the cold fluid, the wall temperature at the cold fluid inlet and outlet and the temperature of the steam.

⑧ At the end of the experiment, firstly turn off the steam control valve, then cut off the steam route of the equipment, close the steam generator (completed by the teacher), and cut off the instrument power supply and the master power supply.

(2) Issues Worth Attention

① A certain amount of air or water must be fed into the tube heat exchanger before the steam valve can be opened, and the condensate originally deposited in the steam pipe must be removed before the steam can be let into the tube heat exchanger.

② When steam is started, the steam valve must be opened slowly, so that steam gradually flows into the heat exchanger and the heating goes slowly. The transition from the "cold state" to the "hot state" should take not less than 10 minutes.

③ In the process of operation, the steam pressure is generally controlled under 0.3 MPa (gauge pressure), because at this condition it is easy to control the system.

④ The parameters must be measured in a steady heat transfer state, and the inert gas must be expelled at all times and the readings of the pressure gauge must be closely watched. In general, the thermal stable state should be kept for at least 5 minutes to ensure the reliability of the data gained.

2.2 "Water-vapor" Heat Transfer Coefficient Experiment

2.2.5 Questions to Consider

① What is the influence of the flow direction of the cold fluid and steam on the heat transfer effect in the experiment?

② Are the density values used in calculating the mass or flow rate of the cold fluid the same as the density values used to obtain the Reynolds number? The density of which position do they respectively represent and under what condition should they be calculated?

③ If, In the course of the experiment, the condensate is not removed in time, what is the influence on the experiment result? How to remove the condensed water in time? What is the influence on the α correlation if the steam of different pressure is used to carry out the experiment?

Chapter 3   Typical Mass Transfer Experiment

3.1 "Ethanol-water" Sieve Plate Rectification Experiment

3.1.1 Purpose of the Experiment

① To understand the basic structure of the sieve plate distillation column and its auxiliary equipment and master the basic operation method of the distillation process.

② To learn the method of judging the stability of the system, and grasp the experimental method for determining the concentration of solutions at the top and bottom of the column.

③ To learn the experimental method to determine the overall efficiency and single plate efficiency of the distillation column and study the influence of the reflux ratio on the separation efficiency of the column.

3.1.2 Principle of the Experiment

(1) Overall Column Efficiency E_T

The overall column efficiency, also known as the total plate efficiency, refers to the ratio of the theoretical number of plates necessary to achieve the designated separation effect to the actual number, namely

$$E_T = \frac{N_T - 1}{N_P} \tag{3-1}$$

Wherein, N_T : The theoretical number plates, including distillation stills, required to complete a certain separation task; N_P : the actual number of plates required to complete a certain separation task, such as $N_P = 10$.

The overall column efficiency simply reflects the average efficiency of the plates in the column and illustrates the influence of the structure, physical property and coefficient, and operating conditions of the plates on the separation capability of the column. The theoretical number N_T of plates required in the column can be calculated using the graphic method according to the known balance relationship of the bi-component system, as well as the composition of the distillate from the top and bottom of the column, the reflux ratio R and the heat condition q, etc.

(2) Single Plate Efficiency E_M

The single plate efficiency, also known as the Murphree efficiency, refers to the ratio of the compositional change of the gas or liquid phase before and after a layer of actual plate to that before and after a layer of the theoretical column x_{n-1}. According to the gas phase compositional change, the single plate efficiency can be expressed as:

$$E_{MV} = \frac{y_n - y_{n+1}}{y_n^* - y_{n+1}} \quad (3\text{-}2)$$

According to the liquid phase compositional change, the single plate efficiency can be expressed as:

$$E_{ML} = \frac{x_{n-1} - x_n}{x_{n-1} - x_n^*} \quad (3\text{-}3)$$

Wherein: y_n, y_{n+1}: the gas phase composition after the plates n and $n+1$, molar fraction; x_{n-1}, x_n: the liquid phase composition of after the plates $n-1$ and n, molar fraction; y_n^*: the gas phase composition in balance to x_n, molar fraction; x_n^*: the liquid phase composition in balance to y_n, molar fraction.

(3) Calculating the theoretical number of plates using graphic method N_T

The graphic method is also called the McCabe-Thiele method, or M-T method in short, whose working principle is the same as the plate-by-plate calculation method, only the plate calculation process is more intuitively demonstrated on the x-y map.

The operating line equation of the distillation section is as follows:

$$y_{n+1} = \frac{R}{R+1}x_n + \frac{x_D}{R+1} \quad (3\text{-}4)$$

Wherein, y_{n+1}: Composition of steam rising from the plate $n+1$ of the distillation section, molar fraction; x_n: Composition of liquid flowing from the plate n of the distillation section, molar fraction; x_D: Composition of the distillate from the top of the column, molar fraction; R: reflux ratio under the reflux boiling point.

The operating line equation of the stripping section is as follows:

$$y_{m+1} = \frac{L'}{L' - W}x_m - \frac{Wx_W}{L' - W} \quad (3\text{-}5)$$

Wherein, y_{m+1}: Composition of steam rising from the plate $m+1$ of the stripping section, molar fraction; x_m: Composition of liquid flowing from the plate m of the stripping section, molar fraction; x_W: Composition of the distillate from at the bottom of the column, molar fraction; L': quantity of the liquid flowing from the stripping section, kmol/s; W: flow rate of distillation residual.

The feeding line (q line) equation can be expressed as:

$$y = \frac{q}{q-1}x - \frac{x_F}{q-1} \tag{3-6}$$

Wherein:
$$q = 1 + \frac{c_{pF}(t_S - t_F)}{r_F} \tag{3-7}$$

Wherein: q : Parameter of the thermal state of the feed; r_F : Latent heat of vaporization in the composition of the feed liquid (kJ/kmol); t_S : The bubble point temperature of the feed liquid (℃); t_F : Temperature of feed liquid (℃); c_{pF} : Specific heat capacity of the feed liquid under the average temperature of $(t_S - t_F)/2$ (kJ/(kmol℃)); x_F : Composition of the feed liquid, molar fraction.

Determining the reflux ratio R:
$$R = \frac{L}{D} \tag{3-8}$$

Wherein, L : flow rate of the reflux liquid(kmol/s); D : flow rate of the distillate liquid (kmol/s).

Equation (3-8) applies only to the reflux under the bubble point, but in actual operation, in order to ensure the updraft can be fully condensed, the volume of the cooling water is generally rather large, making the temperature of the reflux liquid usually lower than the bubble point temperature, which is called cold reflux. As shown in Figure 3-1, the liquid out of the total condenser with temperature t_R and flow rate of L flows to the first plate at the top of the column, and since the reflux temperature is lower than the liquid phase temperature at the first plate, part of the rising steam leaving the first plate is condensed into liquid, as a result of which, the actual flow volume in the column will be greater than the reflux volume from outside the column.

The material and heat balance of the first plate:
$$V_1 + L_1 = V_2 + L \tag{3-9}$$
$$V_1 I_{V1} + L_1 I_{L1} = V_2 I_{V2} + L I_L \tag{3-10}$$

After collating and simplifying equations (3-9) and (3-10), the following approximation can be obtained:
$$L_1 \approx L\left[1 + \frac{c_p(t_{1L} - t_R)}{r}\right] \tag{3-11}$$

That is, the actual reflux ratio:
$$R_1 = \frac{L_1}{D} \tag{3-12}$$

$$R_1 = \frac{L\left[1 + \dfrac{c_p(t_{1L} - t_R)}{r}\right]}{D} \tag{3-13}$$

Wherein, V_1, V_2: the mole-flow of the gas phase leaving the first and second plates

3.1 "Ethanol-water" Sieve Plate Rectification Experiment

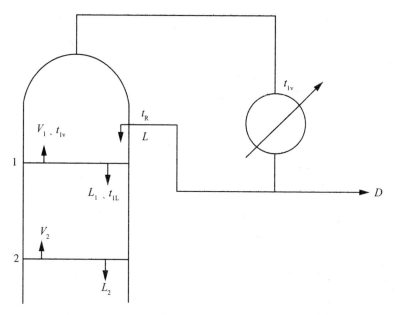

Figure 3-1 Schematic Diagram of the Reflux at the Top of the Column

(kmol/s); L_1: the actual liquid flow rate inside the column(kmol/s); I_{V1}, I_{V2}, I_{L1}, I_L: the enthalpy value corresponding to V_1, V_2, L_1, L (kJ/kmol); r: the latent heat of vaporization in the composition of the reflux liquid (kJ/kmol); c_p: the average specific heat capacity of the reflux liquid under the average temperature of t_{1L} and t_R (kJ/(kmol · ℃)).

1) Full Reflux Operation

In the full reflux operation of rectification, the operation line is the diagonal on the $x-y$ map, as shown in Figure 3-2. By marking steps between the composition at the top and bottom of the column and the operation line and balance line, we can get the theoretical number of plates required.

2) Partial Reflux Operation

In the partial reflux operation, as in Figure 3-3, the main steps of the graphical method are as follows:

① Make the phase equilibrium curve on the $x-y$ map based on material system and operating pressure, and draw the diagonal line as the auxiliary line;

② Determine the three points of $x = x_D$, x_F and x_W on the x axis, and successively draw straight line vertical to the x axis from these three points, which will intersect the diagonal lines at the points of a, f and b.

③ Determine the point c where $y_c = \dfrac{x_D}{R+1}$ on the Y axis, and connect a and c to make the

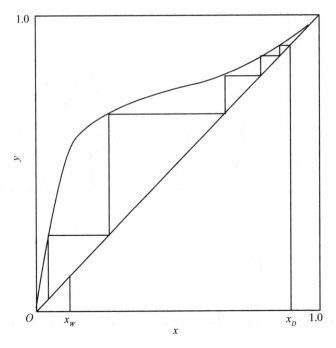

Figure 3-2 Determining the Theoretical Number of Plates in Full Reflux

operation line of the distillation section.

④ Obtain the slope $\dfrac{q}{q-1}$ of the line q from the thermal condition of the feed, and draw the line q from the point f to intersect the operation line of the distillation section at point d;

⑤ Connect the points d and b to make the operation line of the stripping section.

⑥ Draw steps between the balance line and the operation line of the distillation section starting from point a. When the step crosses point d, turn to draw steps between the balance line and the operation line of the stripping section until the step crosses point b.

⑦ The total number of steps drawn is the theoretical number of plates (including reboilers) needed for the whole column. The plate that crosses point d is the feeding plate, and the number of steps above it is the theoretical number of plates needed in the distillation section.

3.1.3 Experimental Devices

The main setup of this experiment is the sieve plate distillation column, supported with a feeding system, reflux system, product outflow pipe, residual liquid discharge pipe, feeding pump and some measuring and controlling instruments. The main structural parameters of the

3.1 "Ethanol-water" Sieve Plate Rectification Experiment

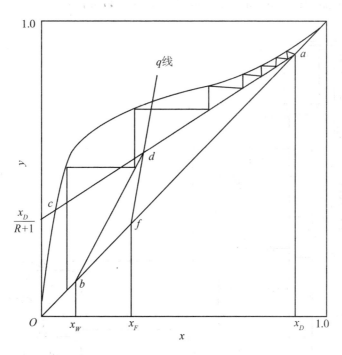

Figure 3-3 Determining the Theoretical Number of Plates in Partial Reflux

sieve plate column: Inner diameter of the column $D = 68$ mm, thickness $\delta = 2$ mm, column sections $\phi 76 \times 4$, plate number $N = 10$, plate spacing $H_T = 100$ mm. The feeding position is the fourth and sixth blocks counting from the bottom. The downcomer is in the bow shape with tooth-shaped weir, the weir length being 56 mm, weir height 7.3 mm, tooth depth 4.6 mm, and tooth number 9. The bottom clearance of the downcomer is 4.5 mm. The diameter of the sieve orifice $d_0 = 1.5$ mm, which is distributed in the equilateral triangular arrangement, the distance between orifices $t = 5$ mm and the number of orifices 74. The column is of the internal electric heating model, the heating power being 2.5 kW and the effective capacity 10 L. The top condenser and bottom heat exchanger of the column are both of the coiled-tube type. The single sampling plate is first and tenth plates counting from bottom, and the upwardly inclined tube is the liquid phase sampling port and the horizontal tube is the gas phase sampling port. The liquid used in this experiment is ethanol. The liquid at the bottom of the column is heated by the electric heater and turned into steam to rise through the plates, and after having mass transfer with the liquid on each plate, enters the shell pass of the coiled-tube type heat exchanger, and after being condensed into liquid, flows out from the liquid collector, part of which flows into the column from the top as reflux liquid, and the rest flows into the product storage tank as distillate; the residual liquid flows into the residual tank through the rotor flowmeter at the bottom of the column. The distillation process is shown in Figure 3-4.

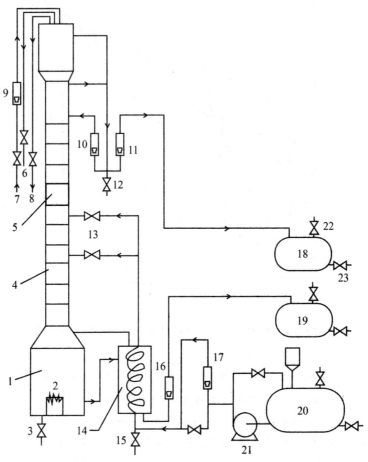

1: bottom of column; 2: electric heater; 3: column bottom liquid discharge port; 4: column section;
5: sight glass; 6: non-condensable gas outlet; 7: cooling water inlet; 8: cooling water outlet;
9: cooling water flowmeter; 10: column top reflux flowmeter; 11: column top effluent flowmeter;
12: column top effluent sampling port; 13: feed valve; 14: heat exchanger; 15: feed liquid sampling port;
16: column bottom residual liquid flowmeter; 17: feed liquid flowmeter; 18: product tank;
19: residual tank; 20: raw material tank; 21: feed pump; 22: emptying valve; 23: liquid discharge valve
Figure 3-4 Schematic Diagram of Experiment Setup of the Sieve Plate Distillation Column

3.1.4 Simulation Devices of the Experiment

① Sieve Plate Distillation Column: The distillation column is of the sieve plate structure, whose body is made of $\phi 57 \times 3.5$mm stainless steel pipe. It has two feed ports and a total of 15 plates, which are made of 1 mm stainless steel and spaced from each other by 10 cm; the orifice rate of the plates is 4%, the orifice diameter 2 mm and the number of orifices 21; the orifices are distributed in the equilateral triangle arrangement; the downcomer is made of $\phi 14 \times 2$ mm stainless steel pipe; the weir height is 10 mm; a WZG-001 micro copper resistance is respectively installed in the top section and the sensitive plate of the column, whose reading is displayed by an XCZ-102 temperature indicator in the instrument cabinet.

② The distiller, $\phi 250 \times 340 \times 3$ mm, is a vertical stainless steel structure and is heated by two 1kW SRY-2-1 type electric heating rods, one of which at constant temperature, the other regulated by an autotransformer, whose reading is displayed by the voltage and ammeter on the instrument cabinet. There are thermometers and pressure gauges on the column to measure the temperature and pressure in it.

③ Condenser: The condenser is of the stainless steel coiler type. The coiler is $\phi 14 \times 2$ mm and 2500 mm long. Tap water is used as the coolant, and the exhaust plug is installed above the condenser.

④ Product tank: The specification of the product tank is $\phi 250 \times 340 \times 3$ mm. It is made of stainless steel, and an observation cover is placed above the tank to observe the flow of the product inside.

In this experiment, the feed solution is an ethanol-water solution, in which the ethanol accounts for 20% (in mole percentage). The solution is stored in the storage tank. The pump is used to feed the column and the column is heated by an electric heater, whose voltage is regulated at the console. After the steam from the bottom of the column reaches the top, it is cooled by the cooler installed on the top (set to "Always Open" in the simulation experiment, so there is no need of a switch for the cooling water valve). After being cooled, the condensate enters the liquid storage tank, and the reflux ratio is controlled by both the opening degree of the reflex valve and the valve on the product collection tank. The product enters the product collection tank. The pressure of the column is regulated by a balanced pressure regulating valve (when the pressure is high, the valve can be opened for depressurization, and in general, pressure in the column is kept under 1.2 atm). Figure 3-5 is Simulation Interface of the Sieve Plate Distillation Experiment. Figure 3-6 is the screen of the console, which marks out the designation of various instruments and switches.

3.1.5 Experimental Procedure

(1) Full Reflux

① Add the feed liquid, 10%-20% (by volume) in concentration, to the storage tank, and open the valve on the feed pipeline. Push the feed liquid into the column by the feed pump, and observe the height of the liquid level in the column, and the feed should take up 2/3 of the capacity of the column chamber. During the feeding, the gate valve on the feeding

Chapter 3 Typical Mass Transfer Experiment

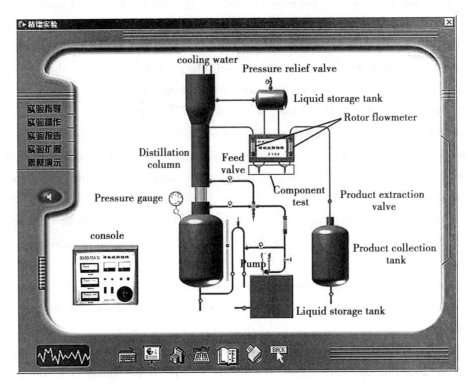

Figure 3-5 Simulation Interface of the Sieve Plate Distillation Experiment

bypass can be opened to speed up the process.

② Close the valve of the inlet pipe on the column, start the power supply of the electric heating tube, gradually increase the heating voltage to let the temperature of the column bottom rise slowly (because the glass part at the center of the column is fragile, if heated too fast, the glass would break easily and shut down the entire system, so the heating process should be as slow as possible).

③ Open the cooling water of the condenser at the top, adjust it to achieve the appropriate condensing amount, and then close outlet pipe at the top so that the whole column is in the full reflux state.

④ When the column top temperature, reflux rate and column bottom temperature are stable, measure the top concentration X_D and bottom concentration X_W to be analyzed with a chromatograph analyzer.

(2) Partial Reflux

① Prepare the ethanol-water solution of certain concentration (10%~20%) in the storage tank.

② When the full reflux operation of the column reaches stability, open the feed valve and

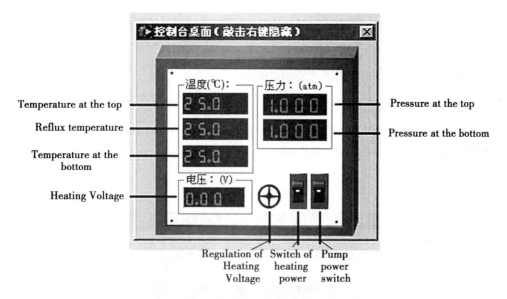

Figure 3-6　Interface of the Console

adjust the feeding rate to achieve the appropriate flow rate.

③ Control the two rotor flowmeters for the reflex and outflow at the top of the column to regulate the reflux ratio R ($R=1$-4).

④ Open the residue flowmeter at the bottom of the column and adjust it to the appropriate flow rate.

⑤Take the sample when the temperature reading and flow rate at the top of and inside the column reach stability.

(3) Sampling and Analysis

① Open the relevant sampling valves at the feeding port, and the top and bottom of the column.

② In plate sampling, use a syringe to slowly draw about 1ml of the sample from the measured plates, and inject it into a clean and dry bottle prepared in advance. Label the bottle cap to avoid mistake and the various samples should be taken in as short a period as possible.

③ Subject the samples to chromatographic analysis.

(5) Issues Worth Attention

① The vent valve at the top of the column must be opened, otherwise the excessive pressure inside would easily lead to danger.

② The feed liquid must be added to 2/3 of the pre-set liquid level before the heating tube is powered, otherwise the excessive low level of liquid in the column will expose the heating wires, damaging them through dry burning.

③ If there is obvious deviation in the temperature of the plates in the experiment, it is because the measured temperature is not of the gas phase, but the mixture of the gas and the liquid instead.

3.1.6 Procedure of Simulation Experiment

① Step one: Full reflux feeding.

a. Open the pump switch. On the console, click the left button of the mouse on the upper end of the pump power switch (the end with the white point) to open the power switch to the pump (see Figure 3-7).

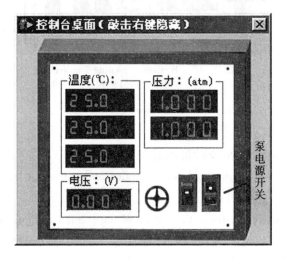

Figure 3-7 Position of the Pump Switch on the Console

b. Open up the feeding pipe, then turn on the valves 1, 2 and 3 in order to feed the liquid to the bottom of the column; when the feed reaches the level of the liquid level gauge marked with the red dot (normal level mark), the feeding is completed, as shown in Figure 3-8.

Figure 3-8 Diagram of Pipeline Position

② Step two:

After the feeding is completed, heat up the bottom of the column, as shown in Figure 3-9. First click on the upper end of the heating power switch to turn on the heater. Click on the heating voltage regulating handle, increasing the voltage by 5V with each left click and decreasing 5V with each right click. Or click the left button of the mouse in the voltage display column, enter the required voltage (0-350V), and then click the left button of the mouse on the blank of the console window to complete the input.

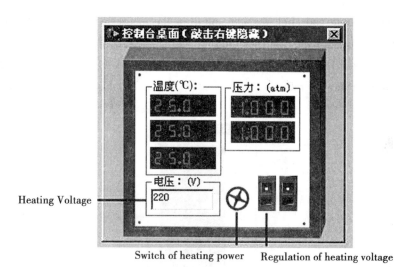

Figure 3-9   Diagram of the Heating Position

③ Step three: Establishing full reflux.

a. After the heating is started and before the beginning of the reflux, note the changes in the temperature of the column bottom and in the pressure at the top of the column. When the pressure at the top of the column is far over one atmospheric pressure (for example by over 0.1 atm), open the pressure relief valve to reduce the pressure (see Figure 3-10). At this time, watch closely the pressure at the top of the column. When it is reduced to one atmospheric pressure, close the valve immediately. Note: After the reflux begins, the pressure relief valve should not be opened again, otherwise the result will be affected.

b. The cooling water at the top of the column is fully open by default, and when the temperature at the bottom of the column reaches about 91℃, condensate begins to appear (there will be fine lines shining between the column top and the storage tanks). At this point, click the left button of the mouse on the rotor flowmeter on the reflux branch, as shown in Figure 3-11. Click the left button of the mouse on the flow regulation knob of the rotor flowmeter

Chapter 3 Typical Mass Transfer Experiment

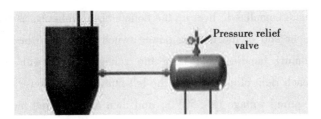

Figure 3-10 Diagram of the Position of the Pressure Relief Valve

to increase the flow rate and click the left button of the mouse to reduce it. Or, the required opening degree (between 0-100, in percentage) can also be filled in the opening display box, and then click the left button of the mouse on the flowmeter. Adjust the opening degree of the valve to 100 to start the full reflux.

Figure 3-11 Schematic Diagram of the Rotor Flowmeter

④ Step four: Reading the full reflux data.

Click the left button of the mouse on "Component Test" to see the content of the component (to be tested with special instrument in practice, omitted here for brevity), as shown in Figure 3-12. When the full reflux goes on for over 10 minutes, the components would be basically stabilized. When the components are stable, click the left button of the mouse on the "Data Processing" in the left-side menu of the main window and fill in the data at the "Original Data" page (Details of the method can be referred to in the standard database operation method). Or, the automatic recording function can be toggled on to enable automatic recording.

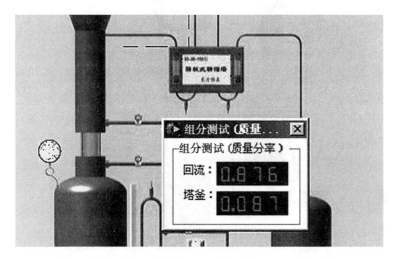

Figure 3-12　Schematic Diagram of the Component Test Data

⑤ Step five: Feed the liquid step by step and begin the partial reflux.

Successively open the feeding valve at the center of the column and the draining valve and product extraction valve at the bottom of the valve, and pay attention to maintaining material balance, liquid level and reflux ratio in the column (see Figure 3-13).

⑥ Step six: Record the data of the partial reflux and fill in the data for processing by reference to the section of the full reflux. Attention:

a. the solution mixing process is omitted, and the original liquid is directly contained in the raw material tank.

b. The heating power switch is simplified from two to one.

c. After the heating is started and before the beginning of the reflux, note the changes in the temperature of the column bottom and in the pressure at the top of the column. When the pressure at the top of the column is far over one atmospheric pressure (for example, by over 0.1 atm), open the pressure relief valve to reduce the pressure. At this time, watch closely the pressure at the top of the column. When it is reduced to one atmospheric pressure, close the valve immediately. Note: After the reflux begins, the pressure relief valve should not be opened again, otherwise the result will be affected.

d. Regarding product inspection, some schools use hydrometers while others use

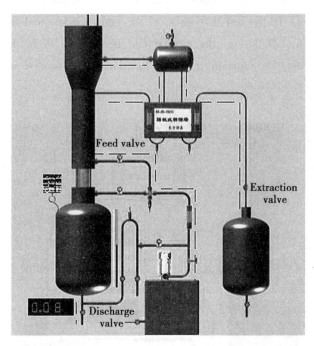

Figure 3-13 Schematic Diagram of Liquid Feeding

refractometer or other instruments. In this simulation experiment, to simplify the matter, the molar fraction of the product is directly given.

3.1.7 Data Recording and Processing

① Fill the original data such as the temperature and composition at the top and bottom of the column, and the readings of the various flowmeters into a table.

② Calculate the theoretical number of plates required using the graphic method for the full reflux and the partial reflux respectively.

③ Calculate the full column efficiency and single plate efficiency.

④ Analyze and discuss the phenomena observed in the experiment.

3.1.8 Recording and Processing of Simulation Experiment

The data processing is basically the same with full reflux and partial reflux. The automatically recorded data (or manually recorded data) can be seen at the Original Data window. The result of the automatic calculation can be seen at the calculation result window, or the result of manual calculation can be filled in the data column (based on which the characteristic curve can be drawn). In the Theoretical Number of Plates tab, the characteristic curve of the distillation column can be drawn from the data in the calculation results (see Figure 3-14).

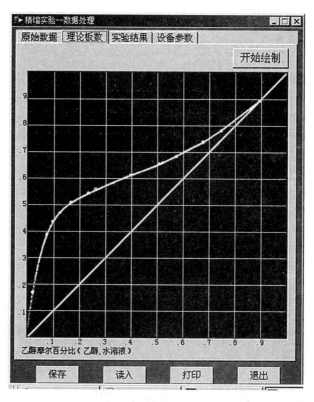

Figure 3-14　Diagram of the Performance Curve

3.1.9　Discussion

① How many parameters should be measured respectively when determining the total plate efficiency and single plate efficiency in the full reflux and partial reflux? Where are the sampling locations?

② How to get x_n^* when the liquid phase composition at the plates n and $n-1$ is measured in full reflux? How to get it in partial reflux?

③ In full reflux, after measuring the liquid phase composition at the plates n and $n-1$, can the single plate efficiency expressed with change in gas phase composition of the plates above the plate n be calculated?

④ What is the value of the qualitative temperature when checking the latent heat of vaporization of the feed fluid?

⑤ If the single plate efficiency is measured to be over 100%, what is the explanation?

Chapter 3 Typical Mass Transfer Experiment

⑥ Try to analyze the reasons for the success or failure of the experiment, and provide suggestions for improvement.

3.2 Filler Absorber

3.2.1 Purpose of the Experiment

① To understand the structure and process of the filler absorber.

② To understand the influence of the change of the inlet conditions of the absorber on the results of the absorption operation.

③ To master the method of determining the gas phase volume overall mass transfer coefficient.

3.2.2 Principle of the Experiment

The two phase mass transfer in the filler column mainly happens on the effective wet surface of the filler, and the height of the filler required to complete a certain absorption task is calculated. Since the equilibrium relation of the gas-liquid phase obeys the Henry's law, that is, the equilibrium line is a straight line, the operation line should also be a straight line.

$$y_1 = \frac{p^*}{p} \tag{3-14}$$

$$\lg p^* = 7.024 - \frac{1161}{224 + T} \tag{3-15}$$

$$G(y_1 - y_2) = L(x_1 - x_2) \tag{3-16}$$

$$\Delta y_m = \frac{y_1 - mx_1 - (y_2 - 0)}{\ln \frac{y_1 - mx_1}{y_2 - 0}} \tag{3-17}$$

$$N_{OG} = \frac{y_1 - y_2}{\Delta y_m} \tag{3-18}$$

$$z = H_{OG} N_{OG} \tag{3-19}$$

$$H_{OG} = \frac{G}{K_{Y_a} \Omega} \tag{3-20}$$

Wherein, y_1: entering molar fraction of the gas phase (bottom); y_2: exiting molar fraction of the gas phase (top); x_1: exiting molar fraction of the liquid phase (bottom); x_2: entering molar fraction of the liquid phase (top); G: molar flow rate of the gas phase (mol/s); L: molar flow rate of the liquid phase (mol/s); z: filler height (m); Δy_m: logarithmic mean of gas phase concentration difference; H_{OG}: height of overall mass transfer unit (m); N_{OG}: number of overall mass transfer unit; m: phase equilibrium constant; K_{Y_a} : gas phase volume overall mass transfer coefficient; Ω: cross-sectional area of the column(m^2); T: liquid phase temperature(K).

3.2.3 Experimental Devices

The air is supplied by an air compressor, whose pressure is set to 0.02MPa through a pressure setting device. After being measured by the rotator flowmeter, it enters the vaporizer containing acetone and produces the acetone-air mixture, and the latter enters the filler column through the bottom. Inside the column, it is exposed to the countercurrent of water sprayed from the top of the column. After most of the acetone content is absorbed by the water, it is discharged from the gas outlet at the top of the column. The water flowing from the bottom of the constant pressure high trough, after being measured by the rotator flowmeter, passes the electric heater, is sprayed into the absorber from the top of the column and after absorbing the acetone in the air, is discharged into the absorption liquid storage tank through the liquid seal device at the bottom of the column (see Figure 3-15). The parameters of absorption column are in Table 3-1, the parameters of the constant pressure tank are in Table 3-2, the parameters of the flowmeter are in Table 3-3.

Table 3-1 **Parameters of Absorption Column**

Diameter (mm)	Column Height (mm)	Type of Filler	Filler Height (mm)	Filler Dimension (mm)
$\phi 41\times 3$	500	China Ranching Ring	400	6×6×1

Table 3-2 **Parameters of the Constant Pressure Tank**

Dimension (mm)	The Insertion Depth of the Suction Pipe in the Trough (mm)
$\phi 300\times 410$	370

Table 3-3 **Parameters of the Flowmeter**

Air Rotor Flowmeter		Liquid Rotor Flowmeter	
Model:	Flow Rate Range(L/h)	Model	Flow Rate Range (L/h)
LZJ-6	100-1000	LZB-4	1-10

3.2.4 Experimental Procedure

① Prepare respectively standard acetone solutions with volume fractions of 0.0%, 1.0%,

Chapter 3 Typical Mass Transfer Experiment

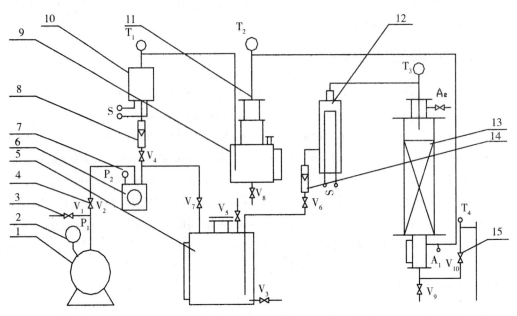

1: air compressor; 2: pressure gauge; 3: air compressor bypass valve; 4: air pressure regulating valve; 5: constant pressure tank for the liquid; 6: pneumatic pressure setting device; 7: pressure gauge; 8: air flowmeter; 9: acetone vaporizer; 10: air heater; 11: acetone vapor-air mixer; 12: water preheater; 13: filler absorption column; 14: rotor flowmeter; 15: liquid seal; T_1、T_2、T_3、T_4: thermometer; V_4、V_6、V_{10}: flow regulating valve; V_3、V_5、V_7、V_8、V_9、V_{11}: switch valve; A_1、A_2: gas inlet and outlet sampling port.

Figure 3-15 Flow Chart of Filler Absorption Experiment

2.0%, 4.0% and 6.0%, measure their refraction rate and draw the standard curve.

② Add the liquid acetone to the acetone vaporizer with a funnel and make the liquid level at over 2/3 of the height of the liquid level meter.

③ Close the outlet valve of the constant pressure trough, send water to it and close the inlet valve before any overflow happens.

④ Start the air compressor, adjust it to make the pressure of the air in the chamber reach 0.05-0.1 MPa, and then adjust the pneumatic pressure regulator to maintain the pressure of the incoming system constant at 0.02 MPa.

⑤ Regulate the rate of the air flow to 40-60 L/h, and the rate of the water flow to 4-8 L/h.

⑥ Keep the rate of the air flow constant for 5 minutes while changing the rate of the water flow, then read the data; keep the rate of the water flow constant for 5 minutes while changing the rate of the air flow, then read the data.

3.2.5 Data Recording and Processing

Draw the standard curve according to Table 3-4 and calculate the molar fraction of the three samples. Water density _____ g/L; Acetone density _____ g/L; Water molecular weight _____ g/mol; Acetone molecular weight _____ g/mol. The data should be recorded in Table 3-5 and Table 3-6.

Table 3-4 **Standard Curve and Determination of the Acetone Content of the Samples**

Sample	Volume Fraction(%)	Molar Fraction(%)	Refraction Rate
Standard Liquid 1			
Standard Liquid 2			
Standard Liquid 3			
Standard Liquid 4			
Standard Liquid 5			
Sample 1			
Sample 2			
Sample 3			

Table 3-5 **Data Recording Table**

No.	Gas Phase Temperature (K)	Liquid Phase Temperature (K)	Q_G(L/h)	Q_L(L/h)	$p_表$(MPa)
1					
2					
3					

Table 3-6 **Data Processing Table**

No.	x_1 %	p^* (kPa)	y_1 %	G (mol/s)	L (mol/s)	y_2 %	m	Δy_m	N_{OG}	H_{OG} (m)	$K_Y a$ (mol/m$^3 \cdot$ s)
1											
2											
3											

3.2.6 Discussion

① In industrial practice, why absorption is carried out under low temperature and high pressure while desorption is carried out under high temperature and atmospheric pressure.

② Try to discuss the relationship between the mass transfer coefficient K_{Ya} of the gas phase and G as wells as L.

3.3 Fluidized Bed Drying Experiment

3.3.1 Purpose of the Experiment

① To understanding the basic structure, technical processes and operation method of the fluidized bed drying device.

② To learn the experimental method for determining the drying properties of materials under constant drying conditions.

③ To grasp the experimental analysis method of calculating the drying rate curve and drying rate at the constant rate stage, the critical water content and balance water content according to the drying curve in the experiment.

④ To study the effect of drying conditions on the characteristics of the drying process.

3.3.2 Principle of the Experiment

When designing the size of a dryer or determining its capacity, the drying property data, such as drying rate, critical moisture content and balance moisture content, etc., of the materials to be dried under a given drying condition are the most fundamental technical parameters for reference. Due to the ever-changing nature of to-be-dried materials in actual production, for most of the specific to-be-dried materials, their drying property data often need to be obtained through experimental measurement. According to whether the air state parameters would change in the drying process, the drying processes can be divided into two categories: the constant drying condition operation and the non-constant drying condition operation. If a large amount of air is used to dry a small amount of material, it can be considered that the temperature and humidity of the wet air are constant during the drying process, and if at the same time the air flow rate and the means of contact between the air and the material are also constant, then this operation is called the constant drying condition operation.

(1) Definition of the Drying Rate

The drying rate is defined as the wet mass removed from a unit dry area (the area providing the wet mass) within a unit time, that is:

$$U = \frac{dW}{Ad\tau} = -\frac{G_c dX}{Ad\tau} \quad \text{kg/(m}^2 \cdot \text{s)} \tag{3-21}$$

Wherein, U: drying rate, also known as drying flux ($kg/(m^2 s)$); A: dry surface area

(m^2); W: wet component of the vapor(kg); τ: drying time(s); G_C: mass of the absolutely dry material(kg); X: moisture content of the material (kg wet/kg dry), the minus symbol meaning X decreases with the progress of the drying time.

(2) Method of Determination of the Drying Rate

① Method 1:

a. Turn on the electronic balance to stand by for use.

b. Turn on the fast moisture meter to stand by for use.

c. Prepare 0.5-1 kg of wet material to stand by for use.

d. Turn on the fan, adjust the air volume to 40-60 m^3/h and turn on the heater to heat it. When the temperature of the hot air is constant (usually set at 70-80℃), add the wet material to the fluidized bed and start timing. At every 4 minutes, take out about 10g of the material and read the temperature of the bed. Place the wet material taken out in the fast moisture meter to obtain the initial mass G_i and final mass G_{iC}, then the instantaneous water content X_i of the material is:

$$X_i = \frac{G_i - G_{iC}}{G_{iC}} \tag{3-22}$$

② Method 2 (optional for digital experimental equipment): Determine the water loss in the drying process by the pressure drop of bed layers.

a. Prepare 0.5-1 kg of wet material to stand by for use.

b. Turn on the fan, adjust the air volume to 40-60 m^3/h and turn on the heater to heat it. When the temperature of the hot air is constant (usually set at 70-80℃), add the wet material to the fluidized bed and start timing. At this time, the pressure difference between the bed layers will decrease with time and last the experiment until the bed layer pressure difference (Δp_e) becomes constant. Then the instantaneous water content X_i of the material is:

$$X_i = \frac{\Delta p - \Delta p_e}{\Delta p_e} \tag{3-23}$$

Wherein, Δp: the pressure difference between the bed layers at the time of τ.

Calculate the instantaneous water content X_i of each moment, and then draw the curse of X_i in relation to the drying time τ_i, as shown in Figure 3-16, which is the drying curve.

The drying curve above can also be converted to obtain the drying rate curve. Calculate the different slope $\frac{dX_i}{d\tau_i}$ under different X_i according to the drying curve determined, then calculate the drying rate U by equation (3-21) and draw the curve of U in relation to X, as shown in Figure 3-17, which is the drying rate curve.

Draw the curve of the temperature of the bed layer in relation to the drying time to obtain the relation curve of the temperature of the bed layer with the drying time.

Chapter 3 Typical Mass Transfer Experiment

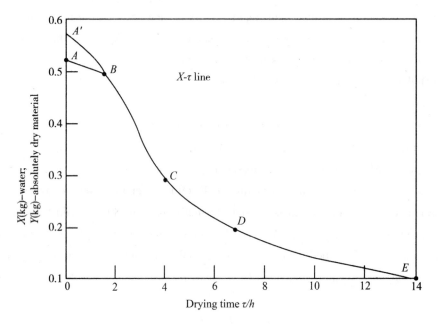

Figure 3-16 Drying Curve Under Constant Drying Conditions

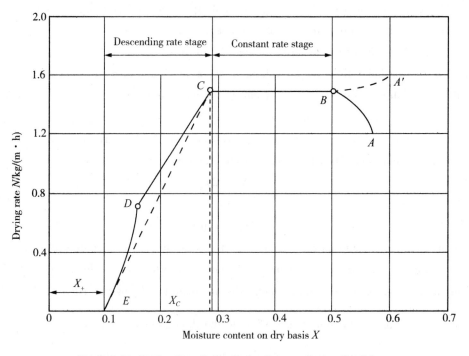

Figure 3-17 Drying Rate Curve Under Constant Drying Conditions

(3) Analysis of the Drying Process

1) Preheating Stage

See the AB or $A'B$ section in Figure 3-16 and Figure 3-17. In the preheating stage, the moisture content of the material drops slightly while the temperature rises to the wet bulb temperature t_W. The drying rate may increase or decrease. The preheating stage is very short and is usually neglected in the drying calculation. In some drying processes there is even no such preheating stage.

2) Constant Rate Drying Stage

The constant rate drying stage is shown as the BC section in Figure 3-16 and Figure 3-17. In this stage, the moisture content of the material is being continuously vaporized, and the water content of the material constantly decreasing. However, because, in this stage, the removed is the unbounded moisture on the surface of the material, the working mechanism of which is the same as that of pure water, thus the surface of the material remains at the wet bulb temperature t_W under constant drying condition, and the mass transfer force remains unchanged, so the drying rate is constant. Thus, in Figure 3-17, the BC section is a horizontal line. As long as the surface of the material is kept sufficiently moist, the drying process of the material is always at a constant speed. The drying rate at this stage depends on the vaporization rate of the water content on the surface of the material, which is also determined by the air drying condition outside the material, as a result of which, this stage is also called the surface vaporization control stage.

3) Decreased Rate Drying Stage

With the progression of the drying process, the gasification rate of the surface moisture exceeds the rate of movement of the internal moister to the surface of the material, local dry patches begin to appear on the surface of the material. Even though at this time, the balance steam pressure of the rest of the surface of the material is the same as that of the saturated vapor pressure of pure water, the drying rate calculated based on the entirety of the outer surface of the material is decreased by the appearance of the dry patches. At this time, the moisture content of the material is called the critical water content, expressed with the symbol Xc, which corresponds to the point C in Figure 3-17 and is called the critical point. After point C, the drying rate is gradually reduced to point D, and the section between C and D is called the first decreased rate stage. When the drying process reaches point D, the entire surface of the material has become dry, and the vaporization surface gradually moves inward, and the heat needed for the vaporization must pass through the solid layer that has been dried to reach the vaporization surface; and the vaporized water from inside the material must also pass through this solid layer to reach the main airstream. The drying rate is reduced because the distance of heat and mass transfer increases. In addition, after point D, the unbound moisture in the material has been removed completely. The next to be vaporized is the various forms of bound water. Therefore, the balance steam vapor pressure will gradually decrease, the mass transfer force decrease, and the drying rate also drop rapidly, until the point E, when the rate will drop to zero. This is called the second decreased rate stage. The shape of the drying rate curve at the decreased rate stage varies with the internal structure of the material, and it does not always show the shape of the curve *CDE* aforementioned. For some porous materials, the boundary between the two stages is not obvious, as if there is only the curve *CD*; for some non-porous absorbent material, vaporization only happens on the surface, the drying rate depends on the diffusion rate of the moisture inside the solid, so the decreased rate stage only has the stage similar to the curve *DE*. As compared with the constant rate stage, in the decreased rate stage, the water content removed from the material is much less, but the drying time required is much longer. In a word, the drying rate in the decreased rate stage depends on the structure, shape and size of the material, but not much on the condition of the drying medium. Therefore, the decreased rate stage is also called the material internal migration control stage.

3.3.3 Experimental Devices

(1) Device Flow

The device flow of this experiment is shown in Figure 3-18.

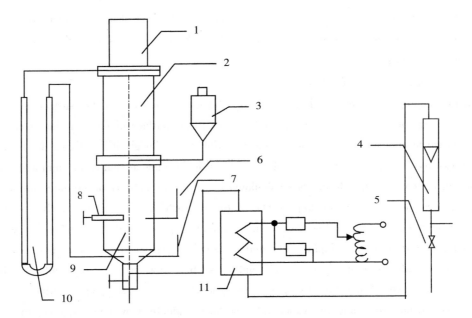

1: dust remover (bag filter), $\phi 130 \times 120$mm; 2: drying column, $\phi 146 \times 8$, high quality glass;
3: water feeder, 0-400mL; 4: gas rotor flowmeter, LZB-25 0-25m^3/n; 5: flow control valve;
6: thermometer, 0-150℃ CU50 copper resistance; 7: thermometer, 0-150℃ CU50 copper resistance;
8: solid material sampler, 2.3g/times; 9: drying material used in the experiment, 30-40 mesh silica gel;
10: differential pressure gauge, ±50cm of mercury; 11: electric heater, 3kW.

Figure 3-18 Schematic Diagram of Experiment Setup and Flow

(2) Main Equipment and Instruments

① Blower: BYF7122, 370 W;

② Electric heater: rated power 2 kW;

③ Drying chamber: ϕ100 mm×750 mm;

④ Drying material: water resistant silica gel;

⑤ Bed layer differential pressure gauge: Sp0014 type differential pressure gauge, or U-shape differential pressure gauge.

3.3.4 Simulation Devices of the Experiment

The screen of the devices of the simulation experiment is shown in Figure 3-19.

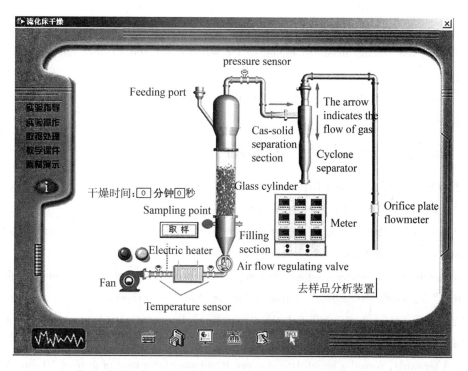

Figure 3-19 Interface of the Devices of the Simulation Experiment

3.3.5 Experimental Procedure

① Connect the gas source and slowly regulate the air flow rate, so that the granular materials in the drying column are in a good fluidization state (note the reading of the differential pressure gauge and let no indicating liquid spill out).

② Add a proper amount of water to the water injector, adjust the copper knob at the bottom of the water injector, so that the flow rate of water in the drying column is not too large. When adding the water, maintain the pulled-out position of the sampler and keep the fluidized state inside the column.

③ Open the air source, open the valve 5, adjust the air flow rate, switch on the power, set the temperature to be between 95-100℃ using the intelligent temperature regulator AI-708.

④ While maintain the flow rate and temperature of the gas, record the temperature of the bed layer at regular interval and sample the solid material to analyze its water content;

⑤ When sampling the solid material, just push in the sampler and immediately pull it out.

⑥The experiment can last until the temperature of the material is obviously increased and the silica gel turns blue.

⑦ Steps to terminate the experiment: cut off the power supply and wait for the gas temperature to drop before cutting the air supply;

⑧ When the column needs to replenish the silicone gel, do so after removing the bag filter.

⑨ When replacing the silica gel, extend the hose of the vacuum cleaner into the column to clean the material out.

3.3.6 Procedure of Simulation Experiment

(1) Process of the Drying Experiment

① Open the fan and start the experiment.

② First adjust the opening of the air flow rate regulating valve to not less than 42, so that the system can enter the fluidized bed stage.

③ Switch on the heater (either manually or automatically) on the instrument cabinet. Click the "Automatic Recording" button to record the "Experimental" data; or manually record the data, when so doing, click the "Sampling" button to sample the materials at the same time.

④ Afterwards, record a set of data at every 10 minutes for a total of at least 10 sets. When the experiment proceeds to a later stage, the sampling interval could be reduced to 6 or 7 minutes. There is time display on the main window. Sampling and recording of experimental data should be carried out in less than one minute for each other.

⑤ The design drying time of this experiment is about 90 to 100 minutes, so the experiment can be stopped after 100 minutes. Proceed to the sample analysis device.

⑥ Samples collected are recorded in the sample selection column, which can be individually selected and weighed. After the weighing, the screen of the electronic scale can display the mass of the sample. Each sample weighs 10 g, which will then be dried and

3.3 Fluidized Bed Drying Experiment

weighed again, so that the net weight of the sample is 10 g minus the water content of the sample. Divide the mass of the water content with 10 g to obtain the moisture content of the sample. The moisture content of the sample can also be manually recorded: click the "Automatic Recording" button in the window to record automatically. With the automatic recording, the moisture content of the sample is first calculated, and then recorded in the corresponding data table.

⑦ After analyzing the sample, click the "Data Processing" button on the main window to enter the data processing interface and select the drying experimental data tab. Pull the horizontal scroll bar to the end, see the drying rate bar, click the "Calculate Drying Rate" button to calculate the drying rate.

⑧ After the original data is completely collected, start drawing the coordinate map.

⑨ Enter the "Critical Moisture Content and Mass Transfer Coefficient" tab, read the value of the critical moisture content from the drying rate curve, and calculate the mass transfer coefficient by clicking the "Calculating Mass Transfer Coefficient" button.

(2) Process of the Fluidization Experiment

① Switch off the heater on the instrument window (turn off all manual and/or automatic switches).

② Close the air flow rate control valve, then slowly open it and properly change the opening of the valve. Record the air flow rate and bed layer pressure drop at each valve opening to the fluidization experiment data tab, which can be recorded manually or automatically. As when the pressure reaches 1 kPa, the flow rate of air would enter the fluidized bed stage, after which the pressure drop of the bed layer is a fixed value. Therefore, at least 5 to 6 sets of data should be obtained before the flow rate increases to 1 kPa, so as to better fit the curve. In the entire experiment, no less than 10 sets of data should be collected.

③ After the original data is completely recorded, draw the corresponding curve. Click the "Fluidized Curve" tab, and then click the "Automatic Drawing" button to draw the fluidization curve.

3.3.7 Data Recording and Processing

The experiment data should be recorded in Table 3-7, and complete the tasks as following:

① Draw the drying curve (that is, the relation curve of water loss with time).

② Draw the drying rate curve according to the drying curve.

③ Read the critical moisture content of the material.

④ Draw the relationship curve of the bed layer temperature with time.

⑤ Analyze and discuss the results of the experiment.

Experiment conditions: Room temperature _____ ;Room relative humidity _____ ; Air flow rate _____ ;Thermometer temperature of hot air _____ .

Table 3-7　　　　　　　　　　**Data Recording table**

No.	Container Weight	Wet Material and Container Weight	Wet Material Weight	Dry Material and Container Weight	Weight of Water Removed	Bed Layer Temperature ℃/10 Minutes	Intermediate Temperature	Reading of Differential Pressure Gauge
1								
2								
3								
4								
5								
6								
7								
8								
9								
10								
11								
12								
13								

Attention: a data is tested every 10 minutes; the sample is baked in the oven at 120℃ for 1 hours; and the scale is of 1/1000 accuracy.

3.3.8　Discussion

① What are the constant drying conditions? In this experiment, what measures are taken to ensure the drying process goes under the constant drying conditions?

② What are the factors that control the drying rate at the constant rate drying stage? What are the factors that control the drying rate at the decreased drying stage?

③ Why should the fan be started first and then the heater? How does the bed layer temperature change in the experiment? Why? How to judge if the experiment has ended?

④ If the flow rate of hot air is increased, what is the change to the drying rate curve? And how would the drying rate at the constant rate drying stage and the critical moisture content change? Why?

参 考 文 献
(References)

[1] Warren L McCable, Julian C Smith, Peter Harriott. Unit Operations of Chemical Engineering [M]. Seventh Edition. New York: McGraw-Hill, 2007.

[2] 叶向群. 化工原理实验及虚拟仿真(双语)[M]. 北京:化学工业出版社,2017.

[3] 王存文,康顺吉,杜治平,等. 化工原理实验(双语)[M]. 北京:化学工业出版社,2014.

[4] 马江权. 化工原理实验[M]. 第三版. 广州:华东理工大学出版社,2016.

[5] 杨祖荣. 化工原理实验[M]. 第2版. 北京:化学工业出版社,2014.

[6] 程远贵,曹丽淑. 化工原理实验[M]. 成都:四川大学出版社,2013.

[7] 都健,王瑶,王刚. 化工原理实验[M]. 北京:化学工业出版社,2017.

[8] 姚克俭,化工原理实验立体教材[M]. 杭州:浙江大学出版社,2015.